AQA Physics Revision Guide

A LEVEL YEAR 2

Jim Breithaupt

OXFORD UNIVERSITY PRESS

Great Clarendon Street, Oxford, OX2 6DP, United Kingdom

Oxford University Press is a department of the University of Oxford. It furthers the University's objective of excellence in research, scholarship, and education by publishing worldwide. Oxford is a registered trade mark of Oxford University Press in the UK and in certain other countries

© Jim Breithaupt 2017

The moral rights of the authors have been asserted

First published in 2017

All rights reserved. No part of this publication may be reproduced, stored in a retrieval system, or transmitted, in any form or by any means, without the prior permission in writing of Oxford University Press, or as expressly permitted by law, by licence or under terms agreed with the appropriate reprographics rights organization. Enquiries concerning reproduction outside the scope of the above should be sent to the Rights Department, Oxford University Press, at the address above.

You must not circulate this work in any other form and you must impose this same condition on any acquirer

British Library Cataloguing in Publication Data
Data available

978 0 19 835775 9

10 9 8 7 6 5 4 3 2 1

Paper used in the production of this book is a natural, recyclable product made from wood grown in sustainable forests.
The manufacturing process conforms to the environmental regulations of the country of origin.

Acknowledgements

Cover: Ferrantraite / iStock

Artwork by Q2A Media

AS/A Level course structure

This book has been written to support students studying for AQA A Level Physics. It covers the Year 2 sections from the specification. The sections covered are shown in the contents list, which also shows you the page numbers for the main topics within each section.

AS exam

Year 1 content
1. Particles and radiation
2. Waves
3. Mechanics and energy
4. Electricity
5. Skills in AS Physics

Year 2 content
6. Further mechanics and thermal physics
7. Fields
8. Nuclear physics

Plus one option from the following:
- Astrophysics
- Medical physics
- Engineering physics
- Turning points in physics
- Electronics

A level exam

A Level exams will cover content from Year 1 and Year 2 and will be at a higher demand.

Contents

| How to use this book | v |

Section 6 Further mechanics and thermal physics

17 Motion in a circle — 2
- 17.1 Uniform circular motion — 2
- 17.2 Centripetal acceleration — 3
- 17.3 On the road — 4
- 17.4 At the fairground — 4
- Practice questions — 6

18 Simple harmonic motion — 8
- 18.1 Oscillations — 8
- 18.2 The principles of simple harmonic motion — 8
- 18.3 More about sine waves — 9
- 18.4 Applications of simple harmonic motion — 10
- 18.5 Energy and simple harmonic motion — 12
- 18.6 Forced vibrations and resonance — 14
- Practice questions — 15

19 Thermal physics — 16
- 19.1 Internal energy and temperature — 16
- 19.2 Specific heat capacity — 17
- 19.3 Change of state — 18
- Practice questions — 20

20 Gases — 22
- 20.1 The experimental gas laws — 22
- 20.2 The ideal gas law — 23
- 20.3 The kinetic theory of gases — 25
- Practice questions — 28

Section 7 Fields

21 Gravitational fields — 29
- 21.1 Gravitational field strength — 29
- 21.2 Gravitational potential — 30
- 21.3 Newton's law of gravitation — 32
- 21.4 Planetary fields — 33
- 21.5 Satellite motion — 34
- Practice questions — 35

22 Electric fields — 36
- 22.1 Field patterns — 36
- 22.2 Electric field strength — 36
- 22.3 Electric potential — 37
- 22.4 Coulomb's law — 38
- 22.5 Point charges — 38
- 22.6 Comparing electric fields and gravitational fields — 39
- Practice questions — 40

23 Capacitors — 42
- 23.1 Capacitance — 42
- 23.2 Energy stored in a charged capacitor — 42
- 23.3 Charging and discharging a capacitor through a fixed resistor — 42
- 23.4 Dielectrics — 45
- Practice questions — 46

24 Magnetic fields — 48
- 24.1 Current-carrying conductors in a magnetic field — 48
- 24.2 Moving charges in a magnetic field — 50
- 24.3 Charged particles in circular orbits — 50
- Practice questions — 52

25 Electromagnetic induction — 54
- 25.1 Generating electricity — 54
- 25.2 The laws of electromagnetic induction — 55
- 25.3 The alternating current generator — 56
- 25.4 Alternating current and power — 56
- 25.5 Transformers — 57
- Practice questions — 58

Section 8 Nuclear physics

26 Radioactivity — 60
- 26.1 The discovery of the nucleus — 60
- 26.2 The properties of α, β, and γ radiation — 61
- 26.3 More about α, β, and γ radiation — 61
- 26.4 The dangers of radioactivity — 62
- 26.5 Radioactive decay — 63
- 26.6 The theory of radioactive decay — 63
- 26.7 Radioactive isotopes in use — 65
- 26.8 More about decay modes — 66
- 26.9 Nuclear radius — 68
- Practice questions — 69

27 Nuclear energy — 71
- 27.1 Energy and mass — 71
- 27.2 Binding energy — 73
- 27.3 Fission and fusion — 75
- 27.4 The thermal nuclear reactor — 77
- Practice questions — 79

Section 9 Option section summaries

- Astrophysics — 81
- Medical physics — 85
- Engineering physics — 91
- Turning points in physics — 96
- Electronics — 102

Answers to practice questions	109
Answers to summary questions	119
For reference	135

How to use this book

This book contains many different features. Each feature is designed to support and develop the skills you will need for your examinations, as well as foster and stimulate your interest in physics.

Worked example
Step-by-step worked solutions.

Maths skill
A focus on maths skills.

Question and model answer
Sample answers to exam style questions.

Summary questions

1 These are short questions at the end of each topic.

2 They test your understanding of the topic and allow you to apply the knowledge and skills you have acquired.

3 The questions are ramped in order of difficulty. Lower-demand questions have a paler background, with the higher-demand questions having a darker background. Try to attempt every question you can, to help you achieve your best in the exams.

Specification references
→ At the beginning of each topic, there are specification references to allow you to monitor your progress.

Key term
Pulls out key terms for quick reference.

Synoptic link
These highlight how the sections relate to each other. Linking different areas of physics together becomes increasingly important, as many exam questions (particularly at A Level) will require you to bring together your knowledge from different areas.

Revision tip
Prompts to help you with your understanding and revision.

Practice questions at the end of each chapter, including questions that cover practical and maths skills.

17.1 Uniform circular motion

Specification reference: 3.6.1.1

▲ **Figure 1** *In uniform circular motion*

Revision tip

The unit of ω is the *radian per second* (rad s^{-1}).

Remember that $360° = 2\pi$ radians.

An object rotating at a steady rate is said to be in **uniform circular motion**. Consider a point on the perimeter of a wheel of radius r rotating at a steady speed.

- The circumference of the wheel, $C = 2\pi r$.
- The frequency of rotation, $f = \frac{1}{T}$, where T is the time for one rotation.
- The angular speed, ω, of the object = its angular displacement per second
$$= \frac{2\pi}{T} = 2\pi f$$
- The speed, v, of a point on the perimeter $= \frac{\text{circumference}}{\text{time for one rotation}}$
$$= \frac{2\pi r}{T} = 2\pi rf = \omega r$$
- The angular displacement of a point, $\theta = \omega t$ where t is the time taken.

Worked example

The wheels of a skateboard each have a diameter of 60 mm. When the skateboarder travels at a speed of 8.5 m s^{-1}, calculate:
a the frequency of rotation of each wheel
b the angular speed of each wheel
c the angle in radians that the wheel turns through in a distance of 20 m.

a Circumference of wheel $= 2\pi r = \pi \times$ diameter $= \pi \times 0.060$ m $= 0.188$ m

Time for one wheel rotation, $T = \frac{\text{circumference}}{\text{speed}} = \frac{0.188 \text{ m}}{8.5 \text{ ms}^{-1}} = 0.0222$ s

Frequency $f = \frac{1}{T} = \frac{1}{0.0222 \text{ s}} = 45$ Hz (2 s.f.)

b Angular speed $\omega = \frac{2\pi}{T} = 290$ rad s^{-1} (2 s.f.)

c Angle the wheel turns through in 20 m,
$$\theta = \frac{\text{distance}}{\text{circumference}} = \frac{20 \text{ m}}{0.188 \text{ m}} = 106 \text{ rad} = 110 \text{ rad (2 s.f.)}$$

Summary questions

1 The wheels of a car each have a diameter of 0.60 m. When the car travels at a speed of 25 m s^{-1}, calculate:
   a the number of turns made by the wheel in a distance of 100 m *(1 mark)*
   b the frequency of rotation of each wheel *(1 mark)*
   c the angular speed of each wheel. *(1 mark)*

2 The International Space Station (ISS) orbits the Earth once every 93 minutes at a height of about 400 km.
   a Calculate the frequency of rotation of ISS. *(1 mark)*
   b The Earth has a radius of about 6400 km. Estimate the speed of ISS. *(1 mark)*

3 In a gear system, a gear wheel of radius 8.2 mm drives a smaller gear wheel of radius 2.5 mm.
When the larger gear wheel turns at a frequency of 0.95 Hz, calculate:
   a the angular speed of the larger gear wheel *(1 mark)*
   b the frequency of rotation of the smaller gear wheel *(3 marks)*
   c i the angular displacement in degrees of the smaller wheel in 0.020 s *(1 mark)*
     ii the distance moved by the gear tooth of the smaller gear wheel in this time. *(1 mark)*

▲ **Figure 2** *Image of a large gear wheel and a small gear wheel in contact with each other*

17.2 Centripetal acceleration

Specification reference: 3.6.1.1

The direction of the velocity of an object in uniform circular motion changes continually as the object moves along its circular path, see Figure 1. This change in the direction of the velocity is towards the centre of the circle. So, the acceleration of the object is towards the centre of the circle and is called **centripetal acceleration**.

For an object moving at constant speed v in a circle of radius r,

$$\text{its centripetal acceleration, } a = \frac{v^2}{r} = \omega^2 r$$

where ω is the angular speed of the object.

Centripetal force

Any object that moves in uniform circular motion is acted on by a resultant force that always acts towards the centre of the circle. The resultant force causes its centripetal acceleration so is referred to as the **centripetal force** on the object.

For an object of mass m moving at constant speed v along a circular path of radius r, applying Newton's second law for constant mass in the form $F = ma$, gives

$$\text{centripetal force } F = \frac{mv^2}{r} = m\omega^2 r$$

▲ Figure 1 Centripetal acceleration

Revision tip

Remember the angular speed $\omega = v/r$

Synoptic link

In Topic 24.3, Charged particles in circular orbits, you will meet the use of a magnetic field to bend a beam of charged particles (e.g., electrons) in a circular path. The magnetic force on the moving charged particles is the centripetal force.

Revision tip

The centripetal force is at right angles to the direction of the object's velocity. Therefore, no work is done by the centripetal force on an object in uniform circular motion because there is no displacement in the direction of the force. The kinetic energy of the object is therefore constant, so its speed is unchanged.

Summary questions

1. A child of mass 28 kg stands on a playground roundabout at a distance of 1.1 m from its centre when it is turning at a steady rate once every 2.3 s. Calculate:
   a the angular speed of the roundabout (1 mark)
   b the child's speed (1 mark)
   c the centripetal force on the child. (1 mark)

2. Figure 2 shows a luggage conveyor belt at an airport. The radius of curvature of the outer edge of the belt at each end is 2.8 m. A suitcase of mass 22 kg is placed near the outer edge of the conveyor belt when the belt is moving at a speed of 0.78 m s^{-1}. Estimate the centripetal force on the suitcase when it moves round the end of the conveyor belt. (1 mark)

▲ Figure 2 Top view

3. A rotating doorway has a circular floor that turns at a steady rate of once every 12 seconds.
   a Calculate the speed of a person standing at a distance of 2.5 m from the centre of the rotating floor. (1 mark)
   b Calculate the centripetal force on a person of mass 61 kg at this position. (1 mark)

17.3 On the road
17.4 At the fairground

Specification reference: 3.6.1.1

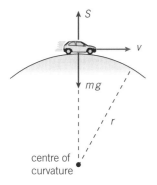

▲ **Figure 1** *Over the top*

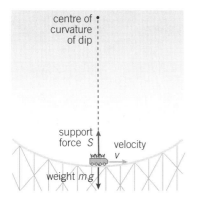

▲ **Figure 2** *In a dip*

> **Revision tip**
> In general, where there are just two forces acting on an object in uniform circular motion, work out the magnitude and direction of the resultant force and equate the magnitude to $\frac{mv^2}{r}$.

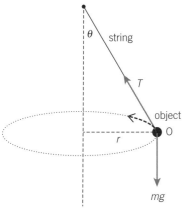

▲ **Figure 3** *The forces acting on an object*

When an object is in uniform circular motion, the forces acting on it give a resultant force which acts as the centripetal force. At A Level, many of the problems you will meet on uniform circular motion involve an object acted on by no more than three forces. The direction of one or more of the forces may change as the object moves round the circle.

Applying Newton's laws of motion

For a given position of an object on its circular path:

1 Identify the forces acting on the object and draw a 'free body diagram' of the object showing the forces acting on it.

2 Resolve the forces into components parallel and perpendicular to the line from the centre of the circle to the object.

3 Equate the resultant of the parallel components towards the centre of the circle to $\frac{mv^2}{r}$.

Using force diagrams

1 A vehicle of mass m moves at speed v on a road that passes over the top of a hill or over the top of a curved bridge.

At the top, the resultant force is directed towards the centre of the circle and is equal to the difference between the weight mg and the support force S of the road on the vehicle. Therefore

$$mg - S = \frac{mv^2}{r}$$

where r is the radius of curvature of the hill.

2 A fairground 'Big Dipper' takes its passengers at high speed through a big dip and 'pushes' them into their seats as they pass through the dip. At the bottom of the dip, the difference between the support force S on a passenger (acting upwards) and their weight mg acts as the centripetal force.

Therefore, for a speed v at the bottom of a dip of radius of curvature r,

$$S - mg = \frac{mv^2}{r}$$

3 A small object O of mass m on the end of a string is swung round in a horizontal circle of radius r at constant speed v. The object is acted on by its weight mg and the tension T in the string (see Figure 3).

The string must be at a non-zero angle θ to the vertical so that the vertical component of T is equal and opposite to the weight. The components of tension parallel and perpendicular to the horizontal are $T\sin\theta$ and $T\cos\theta$ respectively. Therefore:

- the vertical component of tension $T\cos\theta = mg$
- the horizontal component of tension $T\sin\theta = \frac{mv^2}{r}$

Therefore, $\tan\theta = \dfrac{\sin\theta}{\cos\theta} = \dfrac{\left(\frac{mv^2}{r}\right)}{mg} = \dfrac{v^2}{gr}$

Summary questions

$g = 9.81 \, \text{m s}^{-2}$

1. A vehicle of mass 1200 kg passes through a dip in the road at a speed of 22 m s⁻¹. The radius of curvature of the bottom of the dip is 82 m. Calculate:
   a. the centripetal acceleration of the vehicle at the bottom of the dip (*1 mark*)
   b. the magnitude and direction of the support force on the vehicle at the bottom of the dip. (*2 marks*)

2. Figure 4 shows a fairground wheel that rotates at a constant speed in a vertical plane with passengers strapped to the inside of the rim. Each passenger experiences a downward support force when they reach the highest position in the wheel. The wheel takes 7.0 s for each full rotation and it has a radius of 16 m. Calculate the support force on a passenger of mass 48 kg at the highest position. (*4 marks*)

▲ Figure 4

3. A child of mass 23 kg on a swing of length 3.3 m passes through the lowest position of the swing at a speed of 2.8 m s⁻¹. Calculate the support force on the child from the swing when she passes through the lowest position. (*3 marks*)

Synoptic link

The horizontal and vertical components of a force F that is at an angle θ to the vertical must be $F \sin \theta$ and $F \cos \theta$, respectively. See Topic 6.1, Vectors and scalars.

Chapter 17 Practice questions

▲ Figure 1

▲ Figure 2

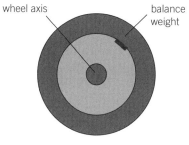

▲ Figure 3

1. The planet Jupiter orbits the Sun once every twelve years on a circular orbit that is about five times larger than the Earth's orbit.

   Which one of the estimates **A–D** gives the following ratio?

   $\dfrac{\text{the speed of Jupiter on its orbit}}{\text{the speed of the Earth on its orbit}}$

   A 0.4   **B** 0.8   **C** 2.4   **D** 5   *(1 mark)*

2. An object O of mass m moves in uniform circular motion at angular speed ω on a circular path of radius r.

   Which one of the expressions **A–D** gives the following ratio?

   $\dfrac{\text{the kinetic energy of the object}}{\text{its centripetal acceleration}}$

   A $\dfrac{m}{2\omega r}$ **B** $\dfrac{m}{2r}$ **C** $\dfrac{mr}{2}$ **D** $\dfrac{m\omega}{2r}$ *(1 mark)*

3. A vehicle of mass m travels at constant speed v over a bridge that has a radius of curvature r.

   Which one of the following expressions **A–D** gives the magnitude of the support force on the vehicle when it is at the top of the bridge?

   A $mg - \dfrac{mv^2}{r}$ **B** $mg + \dfrac{mv^2}{r}$ **C** mg **D** $\dfrac{mv^2}{r} - mg$ *(1 mark)*

4. A spin drier tub of diameter 0.45 m rotates about a horizontal axis at 1400 revolutions per minute (see Figure 2).

   a Calculate:

      i the angular speed of the tub *(1 mark)*

      ii the speed of a point on the wall of the tub *(1 mark)*

      iii the centripetal acceleration of a point on the wall. *(1 mark)*

   b Wet clothing in the tub is forced against the wall of the tub when it spins. The tub is designed with many small holes in the tube wall. Explain why water from the wet clothing leaves the tub through these holes when the tub spins. *(2 marks)*

5. Vehicle wheels are fitted with balance weights to prevent the wheels wobbling at high speed. A balance weight on a wheel is positioned to make the wheel perfectly balanced. A wheel of diameter 0.60 m is fitted with a 25 g balance weight at a distance of 0.21 m from the wheel axis as shown in Figure 3.

   a Calculate the frequency of rotation of the wheel when the wheel is on a vehicle travelling at a speed of 25 m s⁻¹. *(1 mark)*

   b i Calculate the centripetal force F on the balance weight when the wheel is travelling at this speed. *(2 marks)*

      ii Draw a graph to show how the centripetal force F varies with the speed v of the vehicle. *(3 marks)*

6. A cyclist travels at a speed of 8.2 m s⁻¹ on a bicycle that has wheels of diameter 50 cm.

   Calculate:

   a the frequency of rotation of each wheel *(1 mark)*

   b the centripetal acceleration of a point on the tyre. *(1 mark)*

7 A microwave oven has a turntable that rotates at a constant angular speed when the oven is switched on.

  a Describe how you would measure the speed of a point on the edge of the turntable. *(2 marks)*

  b The turntable of a microwave oven has a diameter of 270 mm and it turns at an angular speed of 0.86 rad s^{-1}.

   i Calculate the number of turns made by the turntable in 180 s. *(2 marks)*

   ii Calculate the centripetal acceleration of a small piece of food at the edge of the turntable. *(1 mark)*

8 On a fairground train ride, a train at the highest point of the track travels at a speed of 1.2 m s^{-1}, then descends on the track, and then travels through a dip and onto a level section of track.

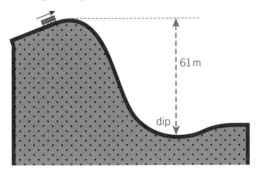

▲ Figure 4

  a The train descends through a vertical height of 61 m before it reaches the bottom of the dip. The total mass of the train and its passengers is 2200 kg.

   i Calculate the loss of potential energy of the train and its passengers in the descent. *(1 mark)*

   ii Calculate the maximum possible speed of the train at the bottom of the descent. *(2 marks)*

  b The dip has a radius of curvature of 52 m. Calculate the force on the train due to the track when the train is at its lowest point and travelling at its maximum possible speed. *(2 marks)*

18.1 Oscillations
18.2 The principles of simple harmonic motion

Specification reference: 3.6.1.2

> **Key terms**
>
> The **displacement** of the object (i.e., distance and direction) from equilibrium continually changes during the motion.
>
> The **amplitude** of the oscillations is the maximum displacement of the oscillating object from equilibrium.
>
> The **frequency**, f, of oscillations is the number of cycles per second made by an oscillating object.
>
> The unit of frequency is the hertz (Hz), which is one cycle per second.
>
> For oscillations of frequency f, the time period $T = \frac{1}{f}$.

> **Revision tip**
>
> Note that 2π radians = 360°. So the phase difference in degrees is $360 \times \frac{\Delta t}{T}$.

> **Revision tip**
>
> 1. The time period is independent of the amplitude of the oscillations.
> 2. Maximum displacement $x_{max} = \pm A$, where A is the amplitude of the oscillations. Therefore, when
> $x_{max} = +A$, the acceleration $a = -\omega^2 A$
> $x_{max} = -A$, the acceleration $a = +\omega^2 A$
> 3. The acceleration equation may also be written as $a = -(2\pi f)^2 x$, where frequency $f = \frac{1}{T}$

The **time period**, T, of the oscillating motion is the time for one complete cycle of oscillation. One full cycle after passing through any position, the object passes through that same position in the same direction.

The **angular frequency** ω of the oscillating motion is defined as $\frac{2\pi}{T} (= 2\pi f)$.
The unit of ω is the radian per second (rad s⁻¹).

Simple harmonic motion is defined as oscillating motion in which the acceleration a is:

1. proportional to the displacement x
2. always in the opposite direction to the displacement x.

The defining equation for simple harmonic motion is

$$\text{acceleration, } a = -\omega^2 x$$

Phase difference

In general, for two objects oscillating at the same frequency,

$$\text{their phase difference in radians} = \frac{2\pi \Delta t}{T}$$

where Δt is the time between successive instants when the two objects are at maximum displacement in the same direction.

For any object undergoing simple harmonic motion, Figure 1 shows how its displacement, x, velocity, v, and acceleration, a, vary with time. All three graphs have the same sinusoidal (sine wave) shape but the velocity is $\frac{1}{4}$ of a cycle ahead of the displacement and the acceleration is $\frac{1}{4}$ of a cycle ahead of the velocity. You can see this from the position of the three peaks at $\frac{T}{2}, \frac{3T}{2}$, and T.

The variation of velocity with time is given by the gradient of the displacement–time graph.

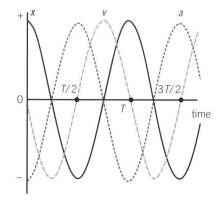

▲ **Figure 1** Displacement (x), velocity, (v) and acceleration (a) against time graphs

The variation of acceleration with time is given by the gradient of the velocity–time graph.

> **Summary questions**
>
> 1. An object suspended from the lower end of a vertical spring is displaced downwards from equilibrium and the time taken for 20 oscillations was measured three times as 15.2 s, 15.6 s, and 15.5 s. Calculate:
>   a i its time period (1 mark)
>   ii its frequency of oscillation (1 mark)
>   b the percentage uncertainty in its frequency. (3 marks)
>
> 2. For the oscillations in **Q1**, calculate the acceleration of the object when its displacement is:
>   a +25 mm (2 marks)   b 0 (1 mark)   c −25 mm. (1 mark)

18.3 More about sine waves

Specification reference: 3.6.1.2

For any object oscillating at frequency f in simple harmonic motion, its acceleration a at displacement x is given by

$$a = -\omega^2 x$$

where $\omega = 2\pi f$

The variation of displacement with time depends on the initial displacement and the initial velocity (i.e., the displacement and velocity at time $t = 0$).

If $x = +A$ when $t = 0$ and the object has zero velocity at that instant, then its displacement at a later time t is given by

$$x = A \cos(\omega t)$$

If $x = 0$ when $t = 0$ and the object has zero velocity at that instant, then its displacement at a later time t is given by

$$x = A \sin(\omega t)$$

▲ **Figure 1** Graphical solutions

Question and model answer

A pendulum bob oscillates in simple harmonic motion after being displaced from rest then released. Its initial displacement was 32 mm and its frequency of oscillation is 0.50 Hz. Calculate:

a the angular frequency of the oscillations   **b** the displacement of the object after 1.60 s

a Angular frequency $\omega = 2\pi f = 3.14$ rad s^{-1}
b Its displacement at time t after release, $x = A \cos(\omega t)$ where $A = 32$ mm
   Therefore at $t = 1.60$ s, $x = A \cos(\omega t) = 32$ mm × cos (3.14 rad s^{-1} × 1.60 s)
   = 32 mm × cos (1.88 rad) = −9.7 mm

Revision tip
The time period T does *not* depend on the amplitude of the oscillating motion. For example, the time period of an object oscillating on a spring is the same, regardless of whether the amplitude is large or small.

Revision tip
Remember that ωt must be in radians. So make sure your calculator is in radian mode when you use the equation $x = A \cos(\omega t)$

In this example the time period is 2.0 s so after 0.6 s, the object must have gone through 0.3 of a cycle. So it would be between $\frac{1}{4}$ and $\frac{1}{2}$ of a cycle at 0.6 s. Look at Figure 1 and you will see its displacement is therefore on the opposite side of the equilibrium position to its initial displacement which is why the answer is negative.

Summary questions

1 An object on a spring oscillates along a vertical line with a time period of 2.3 s and a maximum acceleration of 7.2 m s^{-2}. Calculate:
   a its angular frequency (1 mark)
   b its amplitude. (2 marks)

2 The displacement x from equilibrium of the object in **Q1** is given by $x = A \sin(\omega t)$ where t is the time in seconds from when the object was moving vertically upwards through its equilibrium position.
   a Sketch a graph to show how displacement x varies with t for 1 complete cycle. (2 marks)
   b Calculate the displacement at $t = 2.0$ s. (1 mark)

3 The bob of a simple pendulum was displaced from rest by a distance of 55 mm and then released. The pendulum oscillated in simple harmonic motion with a time period of 0.27 s.
   a Calculate the magnitude of the acceleration of the object half a cycle after it was released. (4 marks)
   b Calculate its displacement and its direction of motion 0.20 s after it was released. (1 mark)

9

18.4 Applications of simple harmonic motion

Specification reference: 3.6.1.3

▲ **Figure 1** *The oscillations of a loaded spring*

Revision tip
1. The time period of the oscillations $T = \frac{1}{f} = 2\pi\sqrt{\frac{m}{k}}$. The time period does *not* depend on g. A mass–spring system on the Moon would have the same time period as it would on Earth.
2. The tension in the spring varies from $mg + kA$ (at maximum displacement downwards) to $mg - kA$ (at maximum displacement upwards) where A = amplitude.

For any oscillating object, the resultant force acting on the object acts towards the equilibrium position. The resultant force is described as a *restoring force* because it always acts towards equilibrium. As long as the restoring force is proportional to the displacement from equilibrium, the acceleration is proportional to the displacement and always acts towards equilibrium. Therefore, the object oscillates with simple harmonic motion.

The oscillations of a mass–spring system

For a small object of mass m attached to a spring which obeys Hooke's Law:

- the time period T of its oscillations is given by $T = 2\pi\sqrt{\frac{m}{k}}$ where k is the spring constant
- its acceleration a at displacement x is given by $a = -\omega^2 x$, where $\omega^2 = \frac{k}{m}$.

Worked example

$g = 9.8\,\text{m s}^{-2}$

A spring of natural length 300 mm hangs vertically with its upper end attached to a fixed point. When a small object of mass 0.20 kg is suspended from the lower end of the spring in equilibrium, the spring is stretched to a length of 379 mm. Calculate:

a i the extension of the spring at equilibrium
  ii the spring constant.
b the time period of oscillations that the mass on the spring would have if the mass was to be displaced downwards slightly then released.

a i Extension of spring at equilibrium is $\Delta L_0 = 79\,\text{mm} = 0.079\,\text{m}$
  ii Spring constant $k = \frac{mg}{\Delta L_0} = \frac{0.20 \times 9.8}{0.079} = 25\,\text{N m}^{-1}$

b $T = 2\pi\sqrt{\frac{0.20}{25}} = 0.56\,\text{s}$

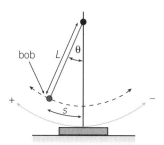

▲ **Figure 2** *The simple pendulum*

Revision tip
1. Note that the time period is independent of the mass of the pendulum bob.
2. Rearranging the time period formula gives frequency
$f = \frac{1}{2\pi}\sqrt{\frac{g}{L}}$

The theory of the simple pendulum

For a simple pendulum that consists of a bob of mass m attached to a thread of length L, as shown in Figure 2, provided angle θ between the thread and the vertical does not exceed approximately 10°:

- the time period T of its oscillations is given by $T = 2\pi\sqrt{\frac{L}{g}}$
- its acceleration a at displacement s is given by $a = -\omega^2 s$, where $\omega^2 = \frac{g}{L}$.

Data analysis

Squaring both sides of the equation $T = 2\pi\sqrt{\frac{m}{k}}$ gives $T^2 = \frac{4\pi^2 m}{k}$ so plotting a graph of T^2 on the y-axis against m on the x-axis should give a straight line through the origin with a gradient of $\frac{4\pi^2}{k}$ in accordance with the general equation for a straight-line graph $y = mx + c$.

You get a similar graph if you square both sides of $T = 2\pi\sqrt{\frac{L}{g}}$ except the gradient of the graph of T^2 against L gives a straight line through the origin with a gradient of $\frac{4\pi^2}{g}$.

Simple harmonic motion

Summary questions

$g = 9.81 \text{ m s}^{-2}$

1. A spring of natural length 300 mm hangs vertically with its upper end attached to a fixed point.
   When a small object of mass 0.20 kg is suspended from the lower end of the spring in equilibrium, the spring is stretched to a length of 379 mm. Calculate:
   - **a** i the extension of the spring at equilibrium (*1 mark*)
     ii the spring constant (*1 mark*)
   - **b** the time period of its oscillations if the object was displaced downwards then released. (*1 mark*)

2. In the arrangement described in **Q1**, the object was replaced by an object of different mass. When the second object was oscillating vertically, its acceleration a at displacement x was given by $a = -250x$.
   - **a** Calculate the time period of the oscillations. (*2 marks*)
   - **b** Calculate the mass of the second object. (*3 marks*)

3. **a** Calculate the time period of a simple pendulum of length 250 mm. (*1 mark*)
   b The length was measured to an accuracy of ± 5 mm. Estimate the percentage uncertainty in the calculated value of the time period. (*2 marks*)

Synoptic link

As the bob passes through equilibrium, the tension T_s acts directly upwards. Therefore, the resultant force on the bob at this instant is

$$T_s - mg = \frac{mv^2}{L}$$

where v is the speed as it passes through equilibrium. See Topic 17.3, On the road.

18.5 Energy and simple harmonic motion

Specification reference: 3.6.1.3

Free oscillations

A freely oscillating object oscillates with a constant amplitude because there is no friction acting on it. The only forces acting on it combine to provide the restoring force. If friction was present, the amplitude of oscillations would gradually decrease, and the oscillations would eventually cease.

The energy of a freely oscillating system changes from kinetic energy to potential energy and back again every half-cycle after passing though equilibrium. As long as friction is absent, the total energy of the system is constant and is equal to its maximum potential energy.

For a system consisting of an object of mass m oscillating on a spring:

- the potential energy E_P at displacement x from equilibrium $= \frac{1}{2}kx^2$, where k is the spring constant of the spring
- the total energy E_T of the system $= \frac{1}{2}kA^2$, where A is the amplitude of the oscillations
- the kinetic energy of the oscillating mass, $E_K = E_T - E_P = \frac{1}{2}k(A^2 - x^2)$.

Energy–displacement graphs

1. The potential energy curve is parabolic in shape, given by $E_P = \frac{1}{2}kx^2$.
2. The kinetic energy curve is an inverted parabola, given by $E_K = E_T - E_P = \frac{1}{2}k(A^2 - x^2)$.
3. The total energy $= \frac{1}{2}kA^2$, which is the potential energy at maximum displacement. The potential energy and kinetic energy curves add together to give a horizontal line for the total energy due to the oscillations.

The simple harmonic motion speed equation

Using $E_K = \frac{1}{2}mv^2$ gives $\frac{1}{2}mv^2 = \frac{1}{2}k(A^2 - x^2)$, where v is the speed of the object at displacement x.

Because $\omega^2 = \frac{k}{m}$, the above equation can be written as $v^2 = \omega^2(A^2 - x^2)$ which gives the speed equation:

$$v = \pm \omega \sqrt{A^2 - x^2}$$

Damped oscillations

In any oscillating system where friction or air resistance is present, the amplitude decreases. The forces dissipate the energy of the system to the surroundings as thermal energy. The motion is said to be **damped** if dissipative forces are present.

Light damping occurs when the time period is independent of the amplitude so each cycle takes the same length of time as the oscillations die away. See Figure 2.

Critical damping is just enough to stop the system oscillating after it has been displaced from equilibrium and released. The oscillating system returns to equilibrium in the shortest possible time without overshooting if the damping is critical.

Heavy damping occurs when the damping is so strong that the displaced object returns to equilibrium much more slowly than if the system is critically damped. No oscillating motion occurs.

Synoptic link

The energy stored in a spring stretched by extension x from its equilibrium length is $\frac{1}{2}kx^2$, where k is the spring constant. See Topic 11.2, Springs.

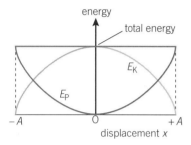

▲ **Figure 1** *Energy variation with displacement*

Revision tip

Making $x = 0$ in the speed equation gives the maximum speed $= \omega A$.

The + and − signs give the velocity direction away (+) or towards (−) equilibrium.

Synoptic link

For a simple pendulum displaced through a height h and released, its maximum kinetic energy = its loss of potential energy (mgh) from its initial position to its lowest position, which is where it passes through equilibrium. See Topic 10.2, Kinetic energy and potential energy.

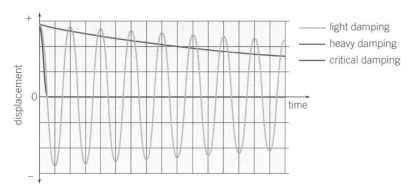

▲ **Figure 2** *Damping*

Summary questions

$g = 9.81 \, \text{m s}^{-2}$

1. A small object of mass 0.150 kg is attached to the lower end of a spring which hangs vertically from a fixed point. The object is displaced vertically downwards from its equilibrium position by a distance of 30 mm and then released. It takes 22.0 s to make 10 complete cycles of oscillation. Calculate:
   a the angular frequency of the oscillations *(1 mark)*
   b the spring constant of the spring *(2 marks)*
   c the maximum speed of the object. *(1 mark)*

2. a Calculate the total energy of the oscillating object in **Q1**. *(1 mark)*
   b Sketch a graph to show how the kinetic energy and the potential energy of the oscillating object in **Q1** varies with displacement. *(4 marks)*

3. A simple pendulum consists of a small metal sphere of mass 0.30 kg attached to a thread. The sphere is displaced from its equilibrium position through a height of 10.0 mm with the thread taut then released. It takes 15.0 s to make 10 complete cycles of oscillation.
   a Calculate the speed of the pendulum bob as it moves through its equilibrium position. *(2 marks)*
   b i Calculate the amplitude A of the oscillations.
      ii Calculate its speed when its displacement is $0.5A$. *(3 marks)*

18.6 Forced vibrations and resonance

Specification reference: 3.6.1.4

▲ **Figure 1** *Resonance curves*

Forced vibrations

The response of an oscillating system to a periodic force depends on the frequency of the periodic force (the applied frequency), see Figure 1.

If the applied frequency is increased gradually from zero:

- **the amplitude of oscillations** of the system increases until it reaches a maximum amplitude at a particular frequency, and then the amplitude decreases again
- **the phase difference between the displacement and the periodic force** increases from zero to $\frac{1}{2}\pi$ at the maximum amplitude, then from $\frac{1}{2}\pi$ to π as the frequency increases further.

Resonance

At the maximum amplitude, the phase difference between the displacement and the periodic force is $\frac{1}{2}\pi$. The periodic force is then exactly in phase with the velocity of the oscillating system, and the system is in **resonance**. The frequency at the maximum amplitude is called the **resonant frequency**.

The lighter the damping:

- the larger the maximum amplitude is at resonance, and
- the closer the resonant frequency is to the natural frequency of the system.

As the applied frequency becomes increasingly larger than the resonant frequency of the oscillating system the amplitude of oscillations decreases more and more.

For an oscillating system with little or no damping, at resonance,

the applied frequency of the periodic force = the natural frequency of the system.

Summary questions

1. A small object of mass m_0 suspended on a vertical spring has a natural frequency of oscillation f_0. The system is subjected to a periodic force of variable frequency f.
   a. Explain why the amplitude of the oscillations is very large when $f = f_0$. *(2 marks)*
   b. When $f = f_0$, the object oscillates with constant energy. Explain why the energy of the object is constant even though a force acts on it. *(2 marks)*

2. With reference to the mass–spring system described in **Q1**, state and explain what the effect would be of replacing:
   a. the object by a small object of mass $4m_0$ *(2 marks)*
   b. replacing the spring with a stiffer spring without changing the object. *(2 marks)*

3. A small object of mass m suspended on a vertical spring is made to resonate by applying a periodic force of frequency 1.8 Hz to it. The spring constant of the spring is $25\ \text{N m}^{-1}$.
   a. Calculate the mass, m, of the object. *(3 marks)*
   b. An identical spring is fitted to the system in parallel with the first spring. Determine the frequency at which the system with the same mass would resonate. *(3 marks)*

Chapter 18 Practice questions

1. A spring fixed at its upper end hangs vertically and supports a small object of mass m attached to its lower end. The spring constant of the spring is k.

   Which of the following expressions **A–D** gives the maximum speed of the object when it oscillates vertically with amplitude A? *(1 mark)*

   A $\dfrac{Ak}{m}$ **B** $\dfrac{A\sqrt{k}}{m}$ **C** $\dfrac{Ak}{\sqrt{m}}$ **D** $\dfrac{A\sqrt{k}}{\sqrt{m}}$

2. Which **one** of the following statements **A–D** about damped oscillations is a correct statement? *(1 mark)*

   A The time period of a lightly damped system increases as the amplitude decreases.

   B A critically damped system never returns to equilibrium after it has been displaced from equilibrium and released.

   C A heavily damped system never oscillates.

   D A damped system subjected to a periodic force can never be made to resonate.

3. Table 1 represents the energy changes of a freely oscillating system during one cycle. Which one of the statements **A–D** is correct? *(1 mark)*

▼ Table 1

	x	E_K	E_P
A	$0 \to +A$	increases	increases
B	$+A \to 0$	increases	decreases
C	$0 \to -A$	decreases	decreases
D	$-A \to 0$	decreases	increases

4. A simple pendulum consists of a small weight on the end of a thread. The weight is displaced from equilibrium and released. It oscillates with an amplitude of 22 mm, taking 35 s to execute 20 oscillations.

   Calculate the magnitude of its maximum acceleration. *(3 marks)*

5. The displacement of an object oscillating in simple harmonic motion varies with time according to the equation $x\,(\text{mm}) = 40\cos 5t$, where t is the time in seconds after the object's displacement was at its maximum positive value.

   a Determine:

      i the amplitude *(1 mark)*

      ii the time period. *(1 mark)*

   b Calculate the displacement of the object at $t = 0.90$ s. *(1 mark)*

6. The upper end of a vertical spring of natural length 250 mm is attached to a fixed point. When a small object of mass 0.15 kg is attached to the lower end of the spring, the spring stretches to an equilibrium length of 320 mm. The object is then displaced vertically from its equilibrium position and released. Calculate the time period of its oscillations. *(3 marks)*

7. a Explain with the aid of a suitable graph what is meant by lightly damped oscillations. *(2 marks)*

   b The amplitude of an oscillating pendulum decreases to 45% of its initial amplitude over five cycles. Estimate its percentage loss of energy per cycle. *(3 marks)*

8. A two-wheeled car trailer has a carrier box of mass 150 kg mounted on two identical springs, one near each wheel (see Figure 1).

   a When the box carries a load of 60 kg, it oscillates in resonance when the trailer travels at a speed of $3.0\,\text{m s}^{-1}$ over speed bumps spaced 12 m apart.

   Calculate the spring constant of each spring. *(5 marks)*

   b Discuss whether or not this effect will happen when the trailer goes over the same speed bumps after it has been unloaded. *(3 marks)*

▲ **Figure 1** *Rear view of the trailer*

19.1 Internal energy and temperature

Specification references: 3.6.2.1; 3.6.2.2

The internal energy of an object is the energy of its atoms and molecules due to their individual movements and positions. The internal energy of an object due to its temperature is sometimes called **thermal energy**. An object's internal energy can change as a result of:

- energy transfer by heating between the object and its surroundings
- work done on the object by a force (including an electrical force).

The first law of thermodynamics

In general, when work is done on or by an object and/or energy is transferred by heating,

the change of internal energy of the object =
the total energy transfer due to work done and heating

About molecules

In a solid, electrical forces between the atoms and molecules hold them together in an arrangement in which they vibrate randomly about fixed positions.

- The energy supplied to raise the temperature of a solid increases the kinetic energy of the molecules.
- The energy supplied to melt a solid raises the potential energy of the molecules because they break free from each other.

In a liquid, the molecules move about at random in contact with each other. The forces between the molecules are not strong enough to hold the molecules in fixed positions.

- The energy supplied to a liquid to raise its temperature increases the average kinetic energy of the liquid molecules.
- Continued supply of energy gives the molecules enough kinetic energy to break free and move away from each other. The potential energy of the molecules increases because they use the kinetic energy they gained to do work and break away from each other.

In a gas or vapour, the molecules also move about randomly but much further apart on average than in a liquid. Heating a gas or a vapour makes the molecules speed up and so gain kinetic energy.

The internal energy of an object is the sum of the random distribution of the kinetic and potential energies of its molecules.

Temperature and temperature scales

The **Celsius scale** of temperature, in units of °C, is defined in terms of ice point, 0 °C, and steam point, 100 °C.

The **absolute scale** of temperature, in units of kelvins (K), starts at **absolute zero**, 0 K, which is the lowest possible temperature.

temperature in °C = absolute temperature in kelvins − 273.15

> **Revision tip**
> When the 1st Law of Thermodynamics is applied to an object, the directions of the energy transfers (i.e., *to* or *from* the object) are very important because the directions determine whether the overall internal energy of the object increases or decreases.

> **Summary questions**
>
> 1. **a** Explain what is meant by the internal energy of an object. *(1 mark)*
>   **b** When a certain gas is compressed, 400 J of work is done on it and 300 J of energy is transferred by heating from it. What change of its internal energy occurs? *(1 mark)*
>
> 2. Describe how the average kinetic energy and the average potential energy of the molecules of a liquid changes when:
>   **a** the liquid is cooled without changing its state *(1 mark)*
>   **b** the liquid solidifies at its melting point. *(2 marks)*
>
> 3. When a stretched rubber band is suddenly released, it becomes warmer for a short time. Describe how its internal energy changes and explain why it becomes warmer for a short time. *(4 marks)*

19.2 Specific heat capacity

Specification reference: 3.6.2.1

The **specific heat capacity**, c, of a substance is the energy needed to raise the temperature of unit mass of the substance by 1 K without change of state. The unit of c is $J\,kg^{-1}\,K^{-1}$.

To raise the temperature of mass m of a substance from temperature T_1 to temperature T_2, the energy needed

$$\Delta Q = mc(T_2 - T_1)$$

The **heat capacity** of an object is the heat supplied to raise the temperature of the object by 1 K.

Measurement of specific heat capacity

1 **For a liquid**, a known mass of the liquid m_1 in an insulated calorimeter of known mass m_{cal} and known specific heat capacity c_{cal} is heated for a measured time t by a 12 V electrical heater in the liquid as shown in Figure 1. A variable resistor is used as shown in Figure 2 to keep the current constant. The heater current I and pd V are measured using the ammeter and voltmeter. If the temperature rise is ΔT, assuming no heat loss to the surroundings,

the electrical energy supplied =
(IVt)

the energy needed to heat the liquid + the energy needed to heat the calorimeter
($m_1 c_1 \Delta T$) ($m_{cal} c_{cal} \Delta T$)

▲ **Figure 1** *Measurement of the specific heat capacity of a liquid*

The specific heat capacity of the liquid c_1 can be calculated from this equation.

2 **For a solid metal**, a block of the metal of known mass m in an insulated container is used. A 12 V electrical heater is inserted into a hole drilled in the metal and used to heat the metal as explained above. A thermometer inserted into a second hole drilled in the metal is used to measure the temperature rise ΔT. Assuming no heat loss to the surroundings,

$$mc\Delta T = IVt$$

▲ **Figure 2** *Circuit diagram*

The specific heat capacity of the solid c can be calculated from this equation.

Continuous flow heating

In an electric shower, water passing steadily through copper coils is heated by an electrical heater. For mass m of liquid, passing through the heater in time t at a steady flow rate, when the outflow is at a constant temperature, assuming no heat loss to the surroundings:

the electrical energy supplied per second, $IV = \dfrac{mc\Delta T}{t}$

where ΔT is the temperature rise of the water and c is its specific heat capacity.

Revision tip
A temperature change is the same in °C as it is in K. If you are given the initial and final temperatures in °C, just calculate the temperature difference in °C.

Synoptic link
If the volume flow rate is given, you need to know the density of the fluid to calculate the rate of flow of mass $\left(\dfrac{m}{t}\right)$. See Topic 11.1, Density.

Summary questions

Use the data in Table 1 for the following calculations.

▼ **Table 1** *Some specific heat capacities*

Substance	Specific heat capacity/ $J\,kg^{-1}\,K^{-1}$
aluminium	900
copper	390
water	4200

1 Calculate the energy needed to heat an insulated copper tank of mass 20 kg containing 430 kg of water from 15 °C to 60 °C. *(2 marks)*

2 Estimate the maximum possible temperature change of an insulated aluminium object that falls to Earth from a height of 8000 m. *(2 marks)*

3 A 5.0 kW electric shower is used to heat the water flowing through it when the flow rate is $3.8 \times 10^{-5}\,m^3\,s^{-1}$. Calculate the maximum possible temperature rise of the water. Density of water = $1000\,kg\,m^{-3}$. *(3 marks)*

19.3 Change of state

Specification reference: 3.6.2.1

The three physical states of a substance, solid, liquid, and vapour, have different physical properties. For example:

- The density of a gas is much less than the density of the same substance in the liquid or the solid state.
- Liquids and gases can flow, but solids can't.

Latent heat

When a solid or a liquid is heated so that its temperature increases, its molecules gain kinetic energy. In a solid, the atoms vibrate mainly about their mean positions. In a liquid, the molecules move about faster, still keeping in contact with each other, but free to move about.

1 **When a solid is heated at its melting point**, its atoms vibrate so much that they break free from each other. The solid therefore becomes a liquid due to energy being supplied at the melting point. The energy needed to melt a solid at its melting point is called **latent heat of fusion**.

2 **When a liquid is heated at its boiling point**, the molecules gain enough kinetic energy to overcome the bonds that hold them close together. The molecules therefore break away from each other to form bubbles of vapour in the liquid. The energy needed to vaporise a liquid is called **latent heat of vaporisation**.

Some solids vaporise directly when heated. This process is called **sublimation**.

For a pure substance, change of state is at constant temperature.

The **specific latent heat of fusion**, l_f, of a substance is the energy needed to change the state of unit mass of the substance from solid to liquid without change of temperature.

The **specific latent heat of vaporisation** of a substance is the energy needed to change the state of unit mass of the substance from liquid to vapour without change of temperature.

So, the energy Q needed to change the state of mass m of a substance from solid to liquid (or liquid to vapour) without change of temperature is given by

$$Q = ml$$

where l is the specific latent heat of fusion (or the specific latent heat of vaporisation). The unit of specific latent heat is $J\,kg^{-1}$.

> **Synoptic link**
>
> If the volume of a substance is given, you need to know its density to find its mass. See Topic 11.1, Density.

Thermal physics

Maths skill: Temperature–time graphs

If a pure solid is heated to its melting point and beyond, its temperature–time graph will be as shown in Figure 1.

▲ **Figure 1** *Temperature against time for a solid being heated*

Assuming that no heat loss occurs during heating, and assuming that energy is transferred to the substance at a constant rate P (i.e., power supplied), then:

- Before the solid melts, $P = mc_s \left(\dfrac{\Delta T}{\Delta t}\right)_s$, where $\left(\dfrac{\Delta T}{\Delta t}\right)_s$ is the rise of temperature per second and c_s is the specific heat capacity of the solid.
  So the rise of temperature per second of the solid is $\left(\dfrac{\Delta T}{\Delta t}\right)_s = \dfrac{P}{mc_s}$.

- After the solid melts, $P = mc_L \left(\dfrac{\Delta T}{\Delta t}\right)_L$, where $\left(\dfrac{\Delta T}{\Delta t}\right)_L$ is the rise of temperature per second and c_L is the specific heat capacity of the liquid.
  So the rise of temperature per second of the liquid is $\left(\dfrac{\Delta T}{\Delta t}\right)_L = \dfrac{P}{mc_L}$.

If the solid has a larger specific heat capacity than the liquid, the rate of temperature rise of the solid is less than that of the liquid. In other words, the liquid heats up faster than the solid.

Summary questions

1. Explain why energy is released when a solid freezes. *(3 marks)*

2. A plastic beaker containing 0.095 kg of water at 12 °C took 600 s to freeze completely. Calculate the energy per second transferred from the water.
   specific heat capacity of water = 4200 J kg^{-1} K^{-1}
   specific latent heat of fusion of water = 3.40×10^5 J kg^{-1} *(3 marks)*

3. **a** In a 3.0 kW electric kettle, 0.015 kg of water is boiled away in 20 s. Calculate the percentage of energy supplied to the kettle that is used to boil the water.
   The specific latent heat of vaporisation of water is 2.25 MJ kg^{-1}. *(2 marks)*

   b A jet of steam at 100 °C was directed into an insulated plastic beaker containing 0.11 kg of water at 20 °C until the water temperature was 55 °C. Calculate the mass of steam that condensed in the beaker. *(5 marks)*

Chapter 19 Practice questions

1 When the outlet of a bicycle pump is blocked and the air in the pump is compressed rapidly, the air becomes warmer for a short time before it returns to its initial temperature. Which **one** of the following changes **A–D** correctly describes the overall energy changes that take place?

	Work done on the air	Energy transfer by heating to surroundings	Internal energy change
A	Yes	No	Increase
B	Yes	Yes	No change
C	Yes	Yes	Decrease
D	No	Yes	No change

(1 mark)

2 Figure 1 shows how the temperature of a pure substance changes as it is heated in an insulated container from below its melting point to above its melting point. The heater transfers energy to the substance at a constant rate.

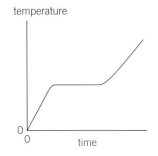

▲ Figure 1

Which **one** of the following statements **A–D** about the substance is correct?

A No energy was transferred to the substance when it was melting.

B The specific heat capacity of the substance in the solid state is less than in the liquid state.

C The rate of change of temperature in the solid state is less than in the liquid state.

D The internal energy of the substance decreased when it was melting.

(1 mark)

3 Which one of the following combinations of SI base units has the same combination of base units as specific latent heat?

A $kg\,m\,s^{-1}$ B $kg\,m\,s^{-2}$ C $m\,s^{-2}$ D $m^2\,s^{-2}$ *(1 mark)*

4 In an experiment to measure the specific heat capacity of an insulated metal block using the method described in Topic 19.2, Specific heat capacity, the following measurements were made:

mass of block = 1.210 ± 0.005 kg
current = 2.60 ± 0.05 A
potential difference = 12.2 ± 0.10 V
heating time = 200 ± 0.40 s
initial temperature = 18.0 ± 0.5 °C
final temperature = 33.5 ± 0.5 °C

a Use this data to calculate the specific heat capacity of the metal.

(3 marks)

b Calculate the percentage uncertainty in:

  i the energy supplied *(2 marks)*

  ii the temperature change *(1 mark)*

  iii the specific heat capacity of the metal. *(1 mark)*

c Discuss how you might test the assumption that no energy is transferred to the surroundings from the metal block during the heating time.

(2 marks)

5 An ice cube at 0 °C is placed in an insulated copper can of mass 0.080 kg containing 0.120 kg of water at 15 °C. After the ice cube has melted and the water has been stirred, the temperature of the water is 12 °C. Calculate the mass of the ice cube.

specific heat capacities: copper = 390 J kg^{-1} K^{-1}, water = 4200 J kg^{-1} K^{-1}

specific latent heat of fusion of water = 3.40×10^5 J kg^{-1} *(6 marks)*

6 In a heating system, hot water is pumped through a steel radiator at a flow rate of 0.042 kg s^{-1}.

▲ Figure 2

   a The water enters the radiator at a temperature of 50 °C and leaves it at a temperature of 44 °C. Calculate the energy transfer from the radiator to the surroundings, assuming the radiator is at a constant temperature.

   The specific heat capacity of water is 4200 J kg^{-1} K^{-1} *(2 marks)*

   b The radiator is made of steel and has a mass of 28 kg. Estimate the time taken for the radiator to warm up from 15 °C, assuming energy is transferred to it at the same rate as in part **a**. In making your estimate, state any further assumptions you have made.

   The specific heat capacity of steel = 560 J kg^{-1} K^{-1} *(4 marks)*

7 In an experiment, a solid was heated by a 50 W electric heater placed inside a hole drilled into the solid. The heater was switched on for seven minutes. The temperature of the solid was measured at the start of every minute. Figure 3 shows how the temperature changed when the solid was heated.

▲ Figure 3

   a Explain why the temperature did not increase at a constant rate. *(3 marks)*

   b The mass of the solid was 0.58 kg. Estimate the specific heat capacity of the solid. *(4 marks)*

   c Describe and explain how the graph would have differed if the solid had been heated for five minutes only. *(2 marks)*

8 When a plastic beaker containing 70 g of water at 10 °C was placed in a freezer, the temperature of the water decreased steadily to 0 °C in 140 s before it started to freeze.

   a Calculate the average rate at which energy was removed from the water.

   The specific heat capacity of water is 4200 J kg K^{-1} *(2 marks)*

   b i Estimate the time taken to freeze the water.

      The specific latent heat of fusion of water is 340 kJ kg^{-1} *(2 marks)*

      ii Discuss the assumptions you made in your estimate and their validity. *(4 marks)*

20.1 The experimental gas laws

Specification reference: 3.6.2.2

> **Key terms**
>
> **Pressure** is defined as the force per unit area exerted normally (i.e., at right angles) on a surface. Pressure is measured in pascals (Pa), where
>
> $1\,\text{Pa} = 1\,\text{N m}^{-2}$
>
> **Absolute zero** is the lowest possible temperature.
>
> **Absolute temperature** T in kelvins (K) = temperature in °C + 273.15

The molecules of a gas move at random with different speeds. The pressure of a gas on a surface is due to the gas molecules hitting the surface. Each impact causes a tiny force on the surface. Because there are a very large number of impacts each second, the overall result is that the gas exerts a measurable pressure on the surface.

- For a fixed mass of gas at constant temperature, $pV = \textbf{constant}$

  where p = gas pressure and V = gas volume.

  A gas that obeys this equation (Boyle's Law) is called an **ideal gas**.

  Figure 1 shows a graph of gas pressure p against volume V for a fixed mass of gas for different constant temperatures.

> **Maths skill: Data analysis**
>
> A graph of pressure on the y-axis against $\dfrac{1}{\text{volume}}$ on the x-axis for a fixed mass of gas at constant temperature gives a straight line through the origin. This is because Boyle's law can be written as $p = \text{constant} \times \dfrac{1}{V}$, which represents the equation $y = mx$ for a straight-line graph through the origin.

- For a fixed mass of an ideal gas at constant pressure, $\dfrac{V}{T} = \textbf{constant}$
  where V is the gas volume and T is the absolute temperature of the gas in kelvins. A graph of V against T is a straight line through the origin.

- For a fixed mass of gas at constant volume, $\dfrac{p}{T} = \textbf{constant}$ where p is its pressure and T is its temperature in kelvins.

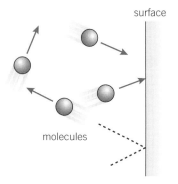

▲ **Figure 1** *Molecules in motion*

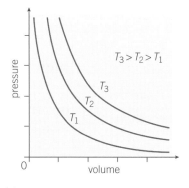

▲ **Figure 2** *Boyle's law*

▲ **Figure 3** *A graph of V against T for an ideal gas*

> **Key terms**
>
> An **isothermal** change is any change at constant temperature.
>
> An **isobaric** change is any change at constant pressure.

> **Revision tip**
>
> For a volume change V of a gas at constant pressure p, the work done by the gas = pV.

Summary questions

1. Air at a pressure of 120 kPa is trapped in a piston and compressed until its volume has decreased from $8.0 \times 10^{-4}\,\text{m}^3$ to $5.0 \times 10^{-4}\,\text{m}^3$ at constant temperature. The gas is then cooled at constant volume from a temperature of 50 °C to 20 °C. Calculate its final pressure. *(3 marks)*

2. A fixed mass of ideal gas at an initial pressure of 100 kPa in a sealed container is cooled from a temperature of 100 °C to −100 °C.
   a. Calculate the pressure of the gas at −100 °C. *(2 marks)*
   b. Sketch a graph to show how the pressure of the gas varied with its temperature in kelvins. *(3 marks)*

3. The volume of a fixed mass of an ideal gas at 15 °C was 0.085 m³. The gas was then heated to 55 °C without change of pressure. Calculate the new volume of this gas. *(2 marks)*

20.2 The ideal gas law

Specification reference: 3.6.2.2; 3.6.2.3

Molecules in a gas

Molecules are too small to see individually. You can see the effect of individual molecules in a gas if you observe smoke particles with a microscope. The smoke particles appear as tiny specks of light wriggling about unpredictably. This type of motion is called **Brownian motion**. Each smoke particle is bombarded unevenly and randomly by individual molecules so it experiences forces due to these impacts, which change its magnitude and direction of motion at random.

Molar mass

One **mole** of a substance consisting of identical particles is defined as the quantity of substance that contains N_A particles. The **molar mass** of a substance is the mass of 1 mol of the substance. The unit of molarity (the number of moles in a given quantity of a substance) is the mol. The unit of molar mass is $kg\,mol^{-1}$.

Therefore, the number of moles n in mass M_S of a substance = $\frac{M_S}{M}$, where M is the molar mass of the substance. For example, because the molar mass of carbon dioxide is 0.044 kg, then n moles of carbon dioxide has a mass of $0.044n$ kg and contains nN_A molecules.

The ideal gas equation

The three experimental gas laws can be combined to give the equation

$$\frac{pV}{T} = \text{constant, for a fixed mass of ideal gas}$$

where p is the pressure, V is the volume, and T is the absolute temperature.

For 1 mole of any ideal gas, the value of $\frac{pV}{T}$ is equal to $8.31\,J\,mol^{-1}\,K^{-1}$ $\left(=\frac{pV}{T}=\frac{101\times 10^3\,Pa \times 0.0224\,m^3}{273\,K}\right)$. This value is called the **molar gas constant**, R.

Therefore for n moles of ideal gas,

$$pV = nRT$$

where V is the volume of the gas at pressure p and temperature T in kelvins.

This equation is called the **ideal gas equation**.

Note: In the equation $pV = nRT$, substituting the number of moles $n = \frac{N}{N_A}$ gives

$$pV = NkT$$

where the **Boltzmann constant** $k = \frac{R}{N_A}$, and N is the number of molecules. Prove for yourself that $k = 1.38 \times 10^{-23}\,J\,K^{-1}$. You will meet k again in the next topic.

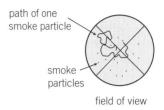

▲ **Figure 1** *Brownian motion*

Synoptic link

Randomness occurs in radioactive decay as well as in Brownian motion of molecules. You can't predict when a random event will take place. See Topic 26.5, Radioactive decay.

Key terms

The Avogadro constant, N_A, is defined as the number of atoms in exactly 12 g of the carbon isotope $^{12}_{6}C$. The value of N_A is 6.023×10^{23} (to four significant figures). So the mass of an atom of $^{12}_{6}C$ is

$$1.993 \times 10^{-23}\,g = \left(\frac{12\,g}{6.023 \times 10^{23}}\right)$$

One atomic mass unit (u) is $\frac{1}{12}$th of the mass of a $^{12}_{6}C$ atom. The mass of a carbon atom is 1.993×10^{-26} kg, so $1\,u = 1.661 \times 10^{-27}$ kg.

Revision tip

Don't mix up molar mass and molecular mass.

Revision tip

The unit of R is the joule per mole per kelvin ($J\,mol^{-1}\,K^{-1}$), which is the same as the unit of

$$\frac{\text{pressure} \times \text{volume}}{\text{absolute temperature} \times \text{number of moles}}$$

This is because the unit of pressure ($Pa = N\,m^{-2}$) × the unit of volume (m^3) is the joule (= 1 N m).

Gases

> **Worked example**
>
> $R = 8.31\ \text{J mol}^{-1}\text{K}^{-1}$
>
> Calculate the number of moles and the mass of air in a 3 litre plastic bottle in which the air pressure 100 kPa, and the temperature of the air in the balloon is 17 °C.
>
> molar mass of air = $0.029\ \text{kg mol}^{-1}$, 1 litre = $0.001\ \text{m}^3$
>
> $T = 273 + 17 = 290\ \text{K}$
>
> Using $pV = nRT$ gives $n = \dfrac{pV}{RT} = \dfrac{100 \times 10^3\ \text{Pa} \times 0.003\ \text{m}^3}{8.31\ \text{J mol}^{-1}\text{K}^{-1} \times 290\ \text{K}} = 0.124\ \text{mol}$
>
> Mass of air = number of moles × molar mass = $0.124\ \text{mol} \times 0.029\ \text{kg mol}^{-1} = 3.6 \times 10^{-3}\ \text{kg}$

Summary questions

$N_A = 6.02 \times 10^{23}\ \text{mol}^{-1}$, $R = 8.31\ \text{J mol}^{-1}\text{K}^{-1}$

1. A cylinder of volume $0.020\ \text{m}^3$ contains nitrogen gas at a pressure of 120 kPa and at a temperature of 20 °C.
   - **a** The molar mass of nitrogen = 0.028 kg. Calculate the mass of gas in the cylinder. *(2 marks)*
   - **b** Calculate the pressure of the gas if the temperature of the gas decreases to 5 °C. *(1 mark)*

2. In a chemistry experiment, $2.2 \times 10^{-5}\ \text{m}^3$ of oxygen gas is collected at a pressure of 101 kPa and a temperature of 12 °C. Calculate the mass of gas collected. *(2 marks)*
   The molar mass of oxygen is $0.032\ \text{kg mol}^{-1}$

3. **a** Estimate the average volume per molecule of air molecules at a pressure of 100 kPa and a temperature of 300 K. *(3 marks)*
   b Use your estimate to make a further estimate of the average spacing between the air molecules in part **a**. *(2 marks)*

20.3 The kinetic theory of gases

Specification reference: 3.6.2.3

The gas laws were explained by assuming that a gas consists of point molecules moving about at random, continually colliding with the container walls. Each impact causes a force on the container. The force of many impacts is the cause of the pressure of the gas on the container walls.

The pressure of a gas can be increased:

- at constant temperature by reducing its volume so the gas molecules travel less distance on average between impacts at the walls. Therefore, there are more impacts per second, and so the pressure is greater.
- at constant volume by increasing its temperature. This raises the kinetic energy and therefore the speed of the molecules. Therefore, the impacts of the molecules on the container walls are harder and more frequent so the pressure is raised.

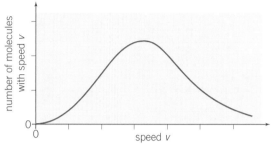

▲ **Figure 1** *Distribution of molecular speeds*

Molecular speeds

The molecules in an ideal gas have a continuous spread of speeds, as shown in Figure 1.

The **root mean square speed** of the molecules,

$$c_{rms} = \left(\frac{c_1^2 + c_2^2 + \ldots + c_N^2}{N}\right)^{\frac{1}{2}}$$

where $c_1, c_2, c_3, \ldots c_N$ represent the speeds of the individual molecules, and N is the number of molecules in the gas.

Revision tip
If the temperature of a gas is raised, the distribution curve becomes flatter and broader because the greater the temperature, the more molecules there are moving at higher speeds.

The kinetic theory equation

For an ideal gas consisting of N identical molecules, each of mass m, in a container of volume V, the pressure p of the gas is given by the equation

$$pV = \frac{1}{3}Nmc_{rms}^2$$

where c_{rms} is the root mean square speed of the gas molecules.

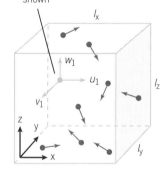

▲ **Figure 2** *Molecules in a box*

Proof

1. **Consider one molecule of mass m in a rectangular box** of dimensions $l_x, l_y,$ and l_z as shown in Figure 2. Let $u_1, v_1,$ and w_1 represent its velocity components in the x, y, and z directions, respectively.

   For each impact of the molecule with the shaded face of the box, its change of momentum = final momentum − initial momentum = $(-mu_1) - (mu_1) = -2mu_1$.

   The time, t, between successive impacts on this face is given by $t = \dfrac{2l_x}{u_1}$.

   Therefore, the force on the molecule = $\dfrac{\text{change of momentum}}{\text{time taken}}$

   $$= \frac{-2mu_1}{\frac{2l_x}{u_1}} = \frac{-mu_1^2}{l_x}$$

   The force F_1 of the impact on the surface is $F_1 = \dfrac{+mu_1^2}{l_x}$ (as it is equal and opposite to the force on the molecule).

   The pressure p_1 of the molecule on the surface = $\dfrac{\text{force } F_1}{\text{area } (l_y l_z)} = \dfrac{mu_1^2}{l_x l_y l_z} = \dfrac{mu_1^2}{V}$

   where V = the volume of the box.

Revision tip
To derive the kinetic theory equation, assume the molecules are point molecules, do not attract each other, move about in continual random motion, collide elastically with each other and with the container surface (i.e., no overall loss of kinetic energy in a collision), and they undergo collisions of much shorter duration than the time between impacts.

Gases

Synoptic link

Recall that the constant k is called the Boltzmann constant. Its value $\left(\frac{R}{N_A}\right)$ is 1.38×10^{-23} J K^{-1}. See Topic 20.2, The ideal gas law.

2 **Consider N identical molecules in the box.**

The total pressure of all the molecules on the surface, $p = p_1 + p_2 + p_3 + \ldots + p_N$ where each subscript refers to each molecule.

Therefore $p = \dfrac{mu_1^2}{V} + \dfrac{mu_2^2}{V} + \dfrac{mu_3^2}{V} + \ldots + \dfrac{mu_N^2}{V} = \dfrac{m}{V}(u_1^2 + u_2^2 + u_3^2 + \ldots + u_N^2) = \dfrac{Nm\bar{u}^2}{V}$

where $\bar{u}^2 = \dfrac{u_1^2 + u_2^2 + u_3^2 + \ldots + u_N^2}{N}$

Because the motion of the molecules is random, there is no preferred direction of motion. Therefore, in terms of the y-components of velocity $v_1, v_2, v_3, \ldots v_N$, or the z-components of velocity $w_1, w_2, w_3, \ldots w_N$:

$p = \dfrac{Nm\bar{v}^2}{V}$ where $\bar{v}^2 = \dfrac{v_1^2 + v_2^2 + v_3^2 + \ldots + v_N^2}{N}$

$p = \dfrac{Nm\bar{w}^2}{V}$ where $\bar{w}^2 = \dfrac{w_1^2 + w_2^2 + w_3^2 + \ldots + w_N^2}{N}$

Therefore, $p = \dfrac{Nm}{3V}(\bar{u}^2 + \bar{v}^2 + \bar{w}^2) = \dfrac{Nm}{3V}c_{rms}^2$ (because $c_{rms}^2 = \bar{u}_{rms}^2 + \bar{v}_{rms}^2 + \bar{w}_{rms}^2$)

So $pV = \dfrac{1}{3}Nmc_{rms}^2$

Molecules and kinetic energy

For an ideal gas, its internal energy is due only to the kinetic energy of the molecules of the gas.

The mean **kinetic energy of a molecule of a gas** =

$\dfrac{\text{total kinetic energy of all the molecules}}{\text{total number of molecules}} = \dfrac{1}{2}mc_{rms}^2$

Since $(pV =) \dfrac{1}{3}Nm(c_{rms})^2 = nRT$

$\dfrac{1}{2}m(c_{rms})^2 = \dfrac{3}{2}\left(\dfrac{nR}{N}\right)T = \dfrac{3}{2}kT$ where $k = \dfrac{nR}{N} = \dfrac{R}{N_A}$

The mean kinetic energy of a molecule of an ideal gas $= \dfrac{3}{2}kT$

The total kinetic energy of n moles of an ideal gas $= N \times \dfrac{3}{2}kT = \dfrac{3}{2}nRT$

and since an ideal gas has no potential energy its

internal energy $= \dfrac{3}{2}nRT$

Worked example

$k = 1.38 \times 10^{-23}$ J mol^{-1} K^{-1}

Calculate the mean kinetic energy of an oxygen molecule in oxygen gas at 0 °C.

$T = 273$ K

Mean kinetic energy $= \dfrac{3}{2}kT$

$= 1.5 \times 1.38 \times 10^{-23}$ J K^{-1} $\times 273$ K

$= 5.7 \times 10^{-21}$ J

Summary questions

$N_A = 6.02 \times 10^{23}$ mol^{-1}, $R = 8.31$ J mol^{-1} K^{-1}, $k = 1.38 \times 10^{-23}$ J K^{-1}

1 Explain in molecular terms why the pressure of a gas in a container increases when its volume is decreased. *(2 marks)*

2 Calculate the root mean square speed of a hydrogen molecule at 20 °C. The molar mass of hydrogen is 0.002 kg mol^{-1} *(3 marks)*

3 A cylinder of volume 0.018 m^3 contains nitrogen gas at a pressure of 120 kPa and a temperature of 323 K. The molar mass of nitrogen is 0.028 kg mol^{-1}. Calculate:
   a the number of moles of nitrogen in the cylinder *(1 mark)*
   b the internal energy of the gas in the container. *(2 marks)*

4 A sample of carbon dioxide is mixed with nitrogen. Calculate the ratio of the root mean square speed of a carbon dioxide molecule to that of a nitrogen molecule in the sample.
nitrogen = 0.028 kg mol^{-1}, carbon dioxide = 0.044 kg mol^{-1} *(4 marks)*

5 An ideal gas of molar mass 0.032 kg mol^{-1} is in a container of volume 0.037 m^3 at a pressure of 100 kPa and a temperature of 300 K. Calculate:
   a the number of moles *(1 mark)*
   b the root mean square speed of the molecules *(3 marks)*
   c the density of the gas. *(2 marks)*

6 The molar mass of air is 0.029 kg m^{-3}.
Calculate the density of air at 15 °C and 101 kPa pressure. *(3 marks)*

Chapter 20 Practice questions

1. A gas cylinder P contains n moles of an ideal gas at pressure p and absolute temperature T. An identical cylinder Q contains the same type of gas at pressure $0.60p$ and a temperature of $0.75T$. Which one of the following options, **A–D**, gives the number of moles of gas in Q?

   A $0.45n$ **B** $0.60n$ **C** $0.75n$ **D** $0.80n$ *(1 mark)*

2. Which one of the following expressions, **A–D**, is a correct expression for the density of an ideal gas of molar mass M at pressure p and absolute temperature T?

   A $\dfrac{p}{MRT}$ **B** $\dfrac{pM}{RT}$ **C** $\dfrac{pRT}{M}$ **D** $\dfrac{MR}{pT}$ *(1 mark)*

3. The root mean square speed of the molecules of a certain gas X of molar mass M at a certain temperature is $800\,\text{m s}^{-1}$. The root mean square speed of the molecules of a different gas Y at the same temperature is $1600\,\text{m s}^{-1}$. Which one of the following alternatives, **A–D**, gives the molar mass of Y?

   A $0.25M$ **B** $0.50M$ **C** $2.0M$ **D** $4.0M$ *(1 mark)*

4. **a** Sketch a graph of pressure against $\dfrac{1}{\text{volume}}$ for a fixed mass of ideal gas at constant temperature. Label this graph O. *(2 marks)*

   b On the *same axes* sketch **two** additional curves labelled P and Q, if the following changes are made:

   i The same mass of gas at a higher constant temperature (P).

   ii A smaller mass of gas at the original constant temperature (Q). *(2 marks)*

5. **a** A cylinder of volume $0.020\,\text{m}^3$ contains an ideal gas at a pressure of $140\,\text{kPa}$ and a temperature of $285\,\text{K}$.

   Calculate:

   i the amount of gas, in moles, in the cylinder

   ii the total kinetic energy of the molecules of gas in the cylinder

   iii the average kinetic energy of a gas molecule in the cylinder. *(4 marks)*

   b The cylinder is heated and the temperature of the gas increases to $320\,\text{K}$. Calculate the increase of internal energy of the gas. *(2 marks)*

6. **a** State what is meant by the root mean square speed of the molecules of a gas. *(1 mark)*

   b i Sketch a graph on the axes in Figure 1 to show the distribution of speeds of the molecules of an ideal gas at constant temperature T_1. *(2 marks)*

   ii Sketch a further graph on the same axes to show the distribution of speeds at a higher temperature T_2. *(2 marks)*

   c Calculate the root mean square speed of nitrogen molecules in air at a temperature of $273\,\text{K}$. The molar mass of nitrogen is $0.028\,\text{kg}$. *(3 marks)*

7. **a** The molecules in a gas move about at random and collide elastically. Explain what is meant by:

   i random motion *(1 mark)* **ii** elastic collisions. *(1 mark)*

   b A gas molecule of mass $5.3 \times 10^{-26}\,\text{kg}$ moving at a speed of $1200\,\text{m s}^{-1}$ collides normally with the wall of the gas container and rebounds without loss of speed.

   i Calculate the change of momentum of this molecule. *(3 marks)*

   ii The pressure of the gas in the container is $120\,\text{kPa}$. Estimate how many such collisions would need to occur each second on each square millimetre of the wall to produce this pressure. *(4 marks)*

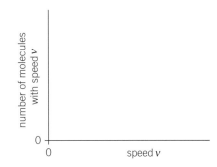

▲ Figure 1

21.1 Gravitational field strength

Specification reference: 3.7.2.2

Any two masses exert a gravitational pull on each other. The mass of an object creates a force field around itself. Any other mass placed in the field is attracted towards the object. The force field round a mass is called a **gravitational field**.

The path which a small mass would follow is called a **field line** or a **line of force**. Figure 1 shows the field lines near a planet. The lines are directed to the centre of the planet because a small object released near the planet would fall towards its centre.

The strength of a gravitational field, g, is the force per unit mass on a small test mass placed in the field.

If a small test mass, m, at a particular position in a gravitational field is acted on by a gravitational force F,

$$g = \frac{F}{m}$$

The unit of gravitational field strength is the newton per kilogram ($N\,kg^{-1}$).

Free fall in a gravitational field

The weight of an object is the force of gravity on it. If an object of mass m is in a gravitational field, the gravitational force on the object is $F = mg$, where g is the gravitational field strength at the object's position. If the object is not acted on by any other force, it accelerates with

$$\text{acceleration } a = \frac{\text{force}}{\text{mass}} = \frac{mg}{m} = g$$

The object therefore falls freely with acceleration g. So, g may also be described as the acceleration of a freely falling object.

Field patterns

1. A **radial field** is where the field lines are always directed towards the same point. See Figure 1.
2. A **uniform field** is where the gravitational field strength has the same magnitude and direction throughout the field. The field lines are therefore parallel to one another and equally spaced.

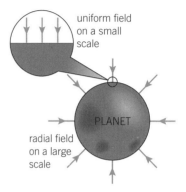

▲ **Figure 1** *Field patterns*

Revision tip
Is the Earth's gravitational field uniform or radial? The force of gravity due to the Earth on a small mass is directed towards the centre of the Earth so the field is therefore radial. However, over distances that are small compared with the Earth's radius, the Earth's gravitational field strength hardly changes so the field may be considered to be uniform.

Summary questions

$g = 9.81\,N\,kg^{-1}$

1. An object of mass 23 kg has a weight of 37 N on the Moon's surface.
   a. Calculate the gravitational field strength of the Moon at its surface. *(1 mark)*
   b. Calculate the weight of space vehicle of mass 250 kg on the Moon's surface. *(1 mark)*

2. a. Explain what is meant by a uniform gravitational field. *(1 mark)*
   b. The gravitational field strength of the Earth is $9.78\,N\,kg^{-1}$ at the equator and $9.83\,N\,kg^{-1}$ at the poles. Calculate the difference in the weight of a 70 kg person at the equator and the same person at the poles. *(1 mark)*

3. The gravitational field strength on the surface of Mars is $3.70\,N\,kg^{-1}$. Calculate the weight on Mars of an astronaut whose weight on the Earth is 650 N. *(2 marks)*

21.2 Gravitational potential

Specification reference: 3.7.2.3

Gravitational potential energy (gpe) is the energy of an object due to its position in a gravitational field. The position for zero gpe is at infinity. In other words, the object would be so far away that the gravitational force on it is negligible. When the object is in the gravitational field, its potential energy is therefore negative.

The gravitational potential, V, at a point in a gravitational field is the work done per unit mass to move a small object from infinity to that point.

Therefore, for a small object of mass m at a position where the gravitational potential is V, the work W that must be done to enable it to escape completely is given by $W = mV$. Rearranging this gives

$$V = \frac{W}{m}$$

The unit of gravitational potential is $J\,kg^{-1}$

Potential gradients

Equipotentials are surfaces of constant potential. Because of this, no work needs to be done to move along an equipotential surface. The equipotentials near the Earth are circles as shown in Figure 1.

At increasing distance from the surface, the equipotentials for equal increases of potential are spaced further apart.

Near the surface over a small region, the equipotentials are horizontal (i.e., parallel to the ground). This is because the gravitational field over a small region is uniform.

The **potential gradient** at a point in a gravitational field is the change of potential per unit distance at that point.

In general, for a change of potential ΔV over a small distance Δr,

$$\text{the potential gradient} = \frac{\Delta V}{\Delta r}$$

Consider a test mass m being moved away from a planet as shown in Figure 2. To move m a small distance Δr in the opposite direction to the gravitational force F_{grav} on it, its gravitational potential energy must be increased:

- by an equal and opposite force F acting through the distance Δr
- by an amount of energy equal to the work done by F, that is, $\Delta W = F\Delta r$.

For the test mass, the change of potential ΔV of the test mass $= \frac{\Delta W}{m}$, so $\Delta V = \frac{F\Delta r}{m}$ and $F = \frac{m\Delta V}{\Delta r}$. As $F_{grav} = -F$, then $F_{grav} = -\frac{m\Delta V}{\Delta r}$. Therefore the gravitational field strength $g = \frac{F_{grav}}{m} = -\frac{\Delta V}{\Delta r}$.

Gravitational field strength $g = -\frac{\Delta V}{\Delta r}$ (equals the negative of the potential gradient) where the minus sign here shows you that g acts in the opposite direction to the potential gradient.

Revision tip
In general, if a small object of mass m is moved from gravitational potential V_1 to gravitational potential V_2, its change of gpe $\Delta E_p = m(V_2 - V_1) = m\Delta V$, where $\Delta V = (V_2 - V_1)$.

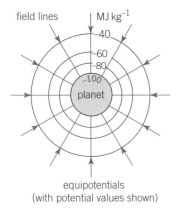

▲ **Figure 1** *Equipotentials near a planet*

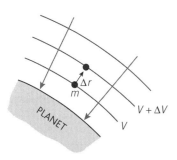

▲ **Figure 2** *Potential gradients*

Revision tip
1 The closer the equipotentials of a gravitational field are, the greater the potential gradient is and the stronger the field is.

2 The potential gradient is always at right angles to the equipotentials, so the direction of the lines of force of the gravitational field are always perpendicular to the equipotentials.

Gravitational fields

Summary questions

$g = 9.81\,\text{N}\,\text{kg}^{-1}$

1. An object of mass 15 kg is lowered through a vertical distance of 50 m.
   a Calculate its change of gravitational potential energy (*1 mark*)
   b Calculate its change of gravitational potential. (*1 mark*)

2. a Define gravitational potential. (*1 mark*)
   b Calculate the change of gravitational potential between a point on the Earth's surface and a point 20.0 m above the surface. (*2 marks*)

3. A rocket of mass 300 kg launched from the Earth's surface reaches a height of 500 km above the surface. After its engine stops it then falls to the ground. The gravitational potential at the Earth's surface is $-62.5\,\text{MJ}\,\text{kg}^{-1}$ and $-57.9\,\text{MJ}\,\text{kg}^{-1}$ at a height of 500 km above the surface.
   a Estimate the mean gravitational field strength between the surface and a point 500 km above it. (*3 marks*)
   b Calculate the change of gravitational potential energy of the rocket when it moves from the surface to 500 km above the surface. (*2 marks*)
   c Calculate the maximum possible speed of the rocket on impact if it falls to the ground from this height. (*3 marks*)

21.3 Newton's law of gravitation

Specification reference: 3.7.2.1

Newton's law of gravitation assumes that the gravitational force between any two *point* objects is:
- always an attractive force
- proportional to the mass of each object
- proportional to $\frac{1}{r^2}$, where r is their distance apart.

In mathematical terms,

$$\text{gravitational force } F = \frac{Gm_1 m_2}{r^2}$$

where m_1 and m_2 are masses of the two objects and G, is the **universal constant of gravitation** (6.67×10^{-11} Nm² kg⁻²).

▲ Figure 1

Revision tip
The distance r is the distance between their centres of mass.

Worked example
$G = 6.67 \times 10^{-11}$ Nm² kg⁻²

The distance from the centre of the Moon to the centre of the Earth is 3.8×10^8 m. The mass of the Moon is 7.4×10^{22} kg, and the mass of the Earth is 6.0×10^{24} kg.
a Calculate the force of gravitational attraction between the Moon and the Earth.
b Calculate the gravitational field strength of the Moon's gravitational field at the Earth.

a $F = \dfrac{6.67 \times 10^{-11} \times 7.4 \times 10^{22} \times 6.0 \times 10^{24}}{(3.8 \times 10^8)^2} = 2.1 \times 10^{20}$ N

b $g = \dfrac{F}{m} = \dfrac{21 \times 10^{20} \text{ N}}{7.4 \times 10^{22} \text{ kg}} = 2.8 \times 10^{-3}$ Nkg⁻¹

Summary questions
$G = 6.67 \times 10^{-11}$ N m² kg⁻²

1. Calculate the force of gravitational attraction between the Moon and a 2000 kg space vehicle in orbit about the Earth when the space vehicle is at a distance of 3.7×10^8 m from the centre of the Moon. The mass of the Moon = 7.4×10^{22} kg. *(1 mark)*

2. The Earth orbits the Sun at a distance of 1.5×10^{11} m from the Sun. Mars orbits the Sun at a distance of 2.3×10^{11} m from the Sun. Calculate the force of gravitational attraction between the Earth and Mars when they are closest to each other.
The mass of the Earth is 6.0×10^{24} kg, the mass of Mars is 6.4×10^{23} kg. *(2 marks)*

3. Estimate the ratio the gravitational force on the Earth due to the Sun and the gravitational force on the Earth due to the Moon. *(3 marks)*
The mass of the Sun = 2.0×10^{30} kg, the distance from the Earth to the Moon is 3.8×10^8 m. See above for mass of the Moon and distance from Sun to Earth.

21.4 Planetary fields

Specification reference: 3.7.2.2; 3.7.2.4

Gravitational field strength

The **magnitude of the gravitational field strength** $g = \dfrac{GM}{r^2}$ at distance r from a point object or from the centre of a sphere of mass M.

For a spherical planet of mass M and radius R, its surface gravitational field strength $g_s = \dfrac{GM}{R^2}$ so $GM = g_s R^2$

Therefore, for distance $r \geq R$, the gravitational field strength $g = \dfrac{GM}{r^2} = \dfrac{g_s R^2}{r^2}$

Gravitational potential near a spherical planet

At or beyond the surface of a spherical planet, the gravitational potential V at distance r from the centre of the planet of mass M is given by

$$V = -\dfrac{GM}{r}$$

The potential energy of an object of mass m at distance r from the centre of the planet $= mV = -\dfrac{GMm}{r}$

The **escape velocity** from a planet is the minimum velocity an object must be given to escape from the planet when projected vertically from the surface.

For an object of mass m to escape from the surface of a planet of radius R to infinity, its initial kinetic energy, $\dfrac{1}{2}mv^2 \geq$ its gain of potential energy $\dfrac{GMm}{R}$

Therefore, its escape velocity $v_{esc} = \sqrt{2gR}$ (as $GM = gR^2$)

Estimating work done

The area under the curve in Figure 3 represents the work done to move the 1 kg mass from infinity to the surface, and is therefore the magnitude of the gravitational potential at the surface. To estimate the work done from the area under the curve, count the number of blocks under the curve. Include blocks that are half full or over as full blocks but not blocks that are less than half full. Use the graph scales to calculate the amount of work corresponding to one full block. The total work done is given by multiplying the number of blocks you counted by the amount of work representing one full block.

▲ Figure 1 Gravitational field strength

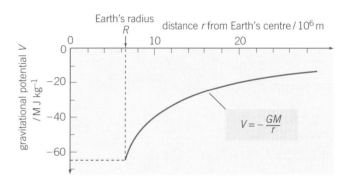

▲ Figure 2 Gravitational potential

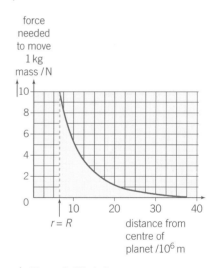

▲ Figure 3 Work done

Revision tip

Know how to draw $\dfrac{1}{r^2}$ and $-\dfrac{1}{r}$ graphs, and note how g changes much more sharply with distance than V does.

Summary questions

$G = 6.67 \times 10^{-11}\,\text{N m}^2\,\text{kg}^{-2}$, $g = 9.81\,\text{N kg}^{-1}$

1. A communications satellite orbits the Earth in a circular orbit of radius 42 000 km.
   a. Calculate the Earth's gravitational field strength at this orbit. The radius of the Earth is 6400 km. *(2 marks)*
   b. i Calculate the Earth's gravitational potential at this orbit.
      ii The satellite has a mass of 200 kg. Calculate its gravitational potential energy. *(3 marks)*

2. Mars has a radius of 3400 km, and its mass is 6.4×10^{23} kg.
   a. Calculate the gravitational field strength at the surface of Mars. *(1 mark)*
   b. Calculate the escape velocity from Mars. *(2 marks)*

21.5 Satellite motion

Specification reference: 3.7.2.4

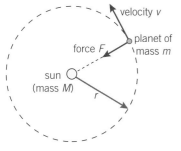

▲ **Figure 1** *Explanation of planetary motion*

For a satellite of mass m orbiting a planet (or a planet orbiting a star) of mass M in a circular orbit of radius r at speed v, the gravitational force $\dfrac{GMm}{r^2}$ = the centripetal force $\dfrac{mv^2}{r}$.

Therefore, the speed of the satellite is given by $v^2 = \dfrac{GM}{r}$.

Since speed $v = \dfrac{\text{circumference of the orbit}}{\text{time period}} = \dfrac{2\pi r}{T}$

$$\dfrac{(2\pi r)^2}{T^2} = \dfrac{GM}{r}$$

Rearranging gives $\dfrac{r^3}{T^2} = \dfrac{GM}{4\pi^2}$

The mass of the Sun

The Earth orbits the Sun once per year on a circular orbit of radius 1.5×10^{11} m. Given $G = 6.67 \times 10^{-11}$ Nm² kg⁻², the above equation can be used to show that the mass of the Sun is 2.0×10^{30} kg.

> **Revision tip**
> For planets orbiting the Sun, because $\dfrac{GM}{4\pi^2}$ is the same for all of the planets, then $\dfrac{r^3}{T^2}$ is the same for all of the planets.

Geostationary satellites

A geostationary satellite orbits the Earth directly above the equator and has a time period of exactly 24 hours. It therefore remains in a fixed position above the equator because it has exactly the same time period as the Earth's rotation. The radius of orbit of a geostationary satellite can be calculated using the equation $\dfrac{r^3}{T^2} = \dfrac{GM}{4\pi^2}$

> **Revision tip**
> A geosynchronous orbit is a 24 hour orbit inclined to the equator (i.e., it crosses the equator).

The energy of an orbiting satellite

Consider a satellite of mass m moving at speed v in a circular orbit of radius r about a spherical planet or star of mass M.

- Its kinetic energy, $E_K = \dfrac{1}{2}mv^2 = \dfrac{1}{2}m \times \dfrac{GM}{r} = \dfrac{GMm}{2r}$

- Its gravitational potential energy, $E_P = mV = -\dfrac{GMm}{r}$

So, the total energy of the satellite $E = E_P + E_K = -\dfrac{GMm}{r} + \dfrac{GMm}{2r} = -\dfrac{GMm}{2r}$

Figure 2 shows how E, E_P, and E_K vary with r. Notice that all three curves are $\dfrac{1}{r}$ curves and that the total energy is always negative.

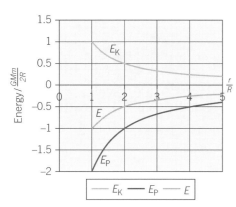

▲ **Figure 2** *Satellite energy*

> **Summary questions**
>
> $G = 6.67 \times 10^{-11}$ Nm² kg⁻², $g = 9.81$ Nkg⁻¹ at the Earth's surface.
> The radius of the Earth is 6400 km and the mass of Earth is 6.0×10^{24} kg.
>
> 1 The International Space Station (ISS) orbits the Earth once every 90 minutes in a circular orbit. Calculate:
>   a the average height of the orbit above the Earth's surface (*3 marks*)
>   b the speed of the satellite. (*3 marks*)
>
> 2 A satellite orbits the Earth in a circular polar orbit at a height of 1200 km. Calculate:
>   a the time period of the orbit (*2 marks*)
>   b the angular speed of the satellite. (*1 mark*)

Chapter 21 Practice questions

$G = 6.67 \times 10^{-11}\,\mathrm{Nm^2\,kg^{-2}}$, $g = 9.81\,\mathrm{Nkg^{-1}}$ at the Earth's surface.

The radius of the Earth is 6400 km and the mass of Earth is 6.0×10^{24} kg.

1 The gravitational field strength at the surface of a spherical planet of radius R is g_s.

Which one of the alternatives **A–D** in Table 1 gives the gravitational field strength and the gravitational potential at height R above the surface?

(1 mark)

2 Which one of the following alternatives **A–D** is the correct combination of SI base units for the potential gradient in a gravitational field?

A ms^{-2} **B** kgs^{-2} **C** m^{-1}s^{-2} **D** kgms^{-2} *(1 mark)*

3 A spherical planet has a radius R and the gravitational potential at its surface is V_s. Which line **A–D** in the graph in Figure 1 shows how the gravitational potential V of its gravitational field varies from its surface with the distance r from the centre of the planet? *(1 mark)*

4 The Moon has a radius of 1740 km and its surface gravitational field strength is $1.62\,\mathrm{Nkg^{-1}}$.

   a Use this data to calculate the mean density of the Moon. *(3 marks)*

   b Calculate the escape velocity from the surface of the Moon. *(2 marks)*

5 Titan is the largest moon of the planet Saturn and it orbits Saturn once every 16.0 days in an orbit that has a mean radius of 1.22×10^9 m.

   a Use this data to calculate the mass of Saturn. *(2 marks)*

   b Saturn is a gaseous sphere which has a radius of 60 300 km. Calculate the mean density of Saturn. *(2 marks)*

   c Titan has a mass of 1.35×10^{23} kg. Calculate the distance from the centre of Titan to the point at which the gravitational field strength of Titan is equal and opposite to the gravitational field strength of Saturn. *(4 marks)*

6 A certain binary star system consists of two identical stars, each of mass M. The two stars move in an orbit of diameter D about the midpoint between their centres (see Figure 2).

   a i Show that for each star, its orbital speed, $v = \sqrt{\dfrac{GM}{2D}}$. *(2 marks)*

   ii Derive an expression for the time period T of each star. *(2 marks)*

   b Two identical stars each of mass 2.0×10^{30} kg orbit each other with time period of 80 years. Calculate the distance between their centres in astronomical units (AU) where 1 AU (the mean distance between the Sun and the Earth) = 1.5×10^{11} m. *(3 marks)*

7 A steel spring has an unstretched length of 300 mm and a spring constant of $5.0\,\mathrm{Nm^{-1}}$. Its limit of proportionality is about 480 mm. Discuss whether or not the spring and a metre ruler could be used to measure the difference between g at the equator and g at the poles, which is about $0.05\,\mathrm{Nkg^{-1}}$. *(4 marks)*

8 Jupiter has a mass of 1.90×10^{27} kg and the mean radius of its orbit about the Sun is 7.8×10^{11} m. The mean radius of the orbit of Mars about the Sun is 2.3×10^{11} m.

   a Estimate the gravitational field strength of Jupiter at Mars when Mars is closest to Jupiter, as shown in Figure 3 *(2 marks)*

   b Calculate the resultant gravitational field strength at Mars due to the Sun and Jupiter when Mars is closest to Jupiter. *(2 marks)*

   The mass of the Sun is 2.0×10^{30} kg.

▼ Table 1

	gravitational field strength	gravitational potential
A	$\frac{1}{2}g_s$	$\frac{1}{2}g_s R$
B	$\frac{1}{2}g_s$	$g_s R$
C	$\frac{1}{4}g_s$	$\frac{1}{2}g_s R$
D	$\frac{1}{4}g_s$	$g_s R$

▲ Figure 1

▲ Figure 2

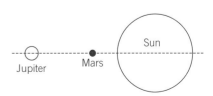

▲ Figure 3

22.1 Field patterns
22.2 Electric field strength

Specification reference: 3.7.3.2

> **Revision tip**
> Like charges repel; unlike charges attract.

▲ Figure 1 *A uniform electric field pattern*

electric field strength, $E = \frac{F}{Q}$ (at Q)

▲ Figure 2 *Electric field strength*

> **Revision tip**
> The unit of E can be written either as the newton per coulomb (NC^{-1}) or as the volt per metre (Vm^{-1}).

> **Synoptic link**
> E is the same everywhere in a uniform electric field, and g is the same everywhere in a uniform gravitational field. See Topic 21.2, Gravitational potential.

Electrical conductors such as metals contain **free electrons**. These are electrons which move about inside the metal and are not attached to any one atom.

Electrically insulating materials do not contain free electrons. All the electrons in an insulator are attached to individual atoms. Some insulators, such as perspex or polythene, are easy to charge because their surface atoms easily gain or lose electrons.

Field lines and patterns

The path a free positive test charge (a small charged object) follows is called a **line of force** or a **field line**. The direction of an electric field line is the direction a positive test charge would move along.

Figure 1 shows the patterns of the field between two oppositely charged parallel plates. The field is *uniform* between the plates because the field lines are parallel to each other.

Electric field strength

The electric field strength, E, at a point in the field is defined as the force per unit charge on a positive test charge placed at that point.

The unit of E is the newton per coulomb (NC^{-1}). Electric field strength is a vector in the same direction as the force on a positive test charge.

If a positive test charge Q at a certain point in an electric field is acted on by force F due to the electric field, the electric field strength $E = \frac{F}{Q}$.

For two parallel plates with a potential difference (pd) V between the plates at separation d, $E = \frac{V}{d}$. This is because the work done moving Q between plates $= Fd = QV$ so $E = \frac{F}{Q} = \frac{V}{d}$.

> **Summary questions**
>
> $e = 1.6 \times 10^{-19} C$
>
> 1. Figure 3 shows some lines of force of the electric field between a negatively charged point object Q and an earthed metal plate.
>   a. Copy Figure 3 and show the direction of the field lines. *(1 mark)*
>   b. State the direction of the force on an electron at:
>   i. point A   ii. point B. *(2 marks)*
>
> 2. A charged metal sphere of radius 0.15 m has a charge of 2.5 nC.
>   a. The charge is distributed evenly on the surface of the sphere. Calculate the number of electrons per mm² on the surface. *(3 marks)*
>   b. When the sphere is discharged, the discharge current is about 0.2 mA. Estimate the discharge time. *(1 mark)*
>
> 3. a. A beam of ions of charge $+3e$ is directed into a uniform electric field of strength 900 kVm⁻¹. Calculate the force on each ion. *(2 marks)*
>   b. The field is due to two oppositely charged parallel plates 50 mm apart. Calculate the pd between the plates. *(2 marks)*

▲ Figure 3

22.3 Electric potential

Specification reference: 3.7.3.3

The electric potential at a certain position in any electric field is the work done per unit positive charge on a positive test charge (i.e., a small positively charged object) when it is moved from infinity to that position. By definition, the position of zero potential energy is infinity

The unit of electric potential is the volt (V), equal to $1\,\text{J}\,\text{C}^{-1}$. For a positive test charge Q placed at a position in an electric field where its electric potential energy is E_P, the electric potential V at this position is given by

$$V = \frac{E_P}{Q}$$

Note:

If a test charge $+Q$ is moved in an electric field from one position where the electric potential is V_1 to another where the electric potential is V_2, the work done $\Delta W = Q(V_2 - V_1)$.

Potential gradients

Equipotentials are surfaces of constant potential. The field lines cross the equipotentials at right angles to the surface.

The **potential gradient** $\dfrac{\Delta V}{\Delta x}$ at any position in an electric field is the change of potential per unit change of distance in a given direction.

At any point in an electric field, the **electric field strength** $E = -\dfrac{\Delta V}{\Delta x}$

The minus sign means that the electric field is in the opposite direction to the potential gradient. See Figure 2

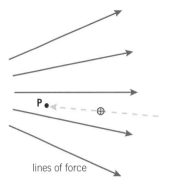

▲ **Figure 1** *Electric potential energy*

Worked example

A $+0.5\,\mu\text{C}$ test charge is moved into an electric field from infinity to reach a certain position P, where the electric potential is $-3000\,\text{V}$. Calculate the electric potential energy of the test charge at P.

$E_P = QV = +0.5 \times 10^{-6}\,\text{C} \times (-3000\,\text{V})$
$\quad = -1.5 \times 10^{-3}\,\text{J}$

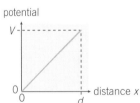

▲ **Figure 2** *A uniform potential gradient*

Summary questions

$e = 1.6 \times 10^{-19}\,\text{C}$

1. An electron in a uniform electric field of strength $8.0\,\text{kV}\,\text{m}^{-1}$ moves from X to Y through a distance of 28 mm in the opposite direction to the direction of the electric field.
    a Calculate the work done by the electric field on the electron. *(2 marks)*
    b The potential at X was $+400\,\text{V}$. Calculate the potential at Y. *(2 marks)*

2. An ion of charge $+2e$ was moved a distance of 36 mm from A where the potential was $+1000\,\text{V}$ to a point B at a higher potential of $6000\,\text{V}$.
    a Calculate the work done by the ion when it moved from A to B. *(2 marks)*
    b Estimate an average value for the potential gradient between A and B. *(2 marks)*

3. Two vertical parallel plates L and M are at a fixed separation of 80 mm. Plate L is at a potential of $+4200\,\text{V}$ and plate M is earthed.
    a Sketch a graph to show how the potential between L and M varies along the shortest straight line between them. *(2 marks)*
    b Calculate: i the potential gradient between L and M
    ii the electric field strength between the plates.
    (3 marks)

37

22.4 Coulomb's law
22.5 Point charges

Specification reference: 3.7.3.1; 3.7.3.3

a unlike charges attract   b like charges repel

$$F = \frac{1}{4\pi\varepsilon_0} \frac{Q_1 Q_2}{r^2}$$

▲ **Figure 1** Coulomb's law

Revision tip
Remember to square r when calculating F.
When substituting charges, get the powers of 10 correct: $\mu = 10^{-6}$, $n = 10^{-9}$, $p = 10^{-12}$.

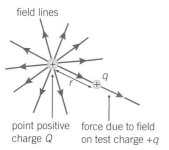

▲ **Figure 2** Force near a point charge Q

Revision tip
A negative E for the field of a point charge indicates the point charge is negative. But a negative V indicates a value less than zero: E is a vector, whilst V is a scalar.

▲ **Figure 3** The electric field and potential near a point charge

Coulomb's law states that the force F between two point charges Q_1 and Q_2 is proportional to the amount of each charge and to the inverse of the square of the distance between the two charges. Coulomb's law can be expressed as the following equation $F = \frac{1}{4\pi\varepsilon_0} \frac{Q_1 Q_2}{r^2}$

where r is the distance between the two charges and $\frac{1}{4\pi\varepsilon_0}$ is the constant of proportionality. See Topic 22.5 of the Student Book if necessary.

The electric field near a point charge

At distance r from a point charge Q,

the electric field strength $E = \frac{1}{4\pi\varepsilon_0} \frac{Q}{r^2}$

the electric potential $V = \frac{1}{4\pi\varepsilon_0} \frac{Q}{r}$

the potential energy of point charges q and Q (at separation r) = $qV = \frac{1}{4\pi\varepsilon_0} \frac{qQ}{r}$

Where two or more point charges are present, the resultant electric field strength is the vector sum of the electric field strengths of the point charges and the overall potential is the sum of the potentials of the point charges.

Graphs

Figure 3 shows how the electric field strength and the electric potential vary with distance from a point charge (or from the centre of a charged metal sphere).

Electric field strength = the negative of the gradient of a potential against distance graph.

Change of potential = area under an electric field strength against distance graph.

Summary questions

$\varepsilon_0 = 8.85 \times 10^{-12}$ Fm^{-1} $e = 1.6 \times 10^{-19}$ C

1. Calculate the force between two identical ions, each of charge $+2e$, at a separation of 5.6×10^{-7} m. *(2 marks)*

2. Two charged point objects, X and Y, at a separation of 1.6×10^{-2} m exert a force on each other of 5.1×10^{-5} N. When the charge of X is increased by 2.1 nC, the force increases to 7.8×10^{-5} N. Calculate the charge of each object. *(5 marks)*

3. A proton and an electron are at a separation of 0.11 nm.
   a. Calculate their potential energy at this separation. *(2 marks)*
   b. Calculate the electric field strength at the midpoint between the proton and electron. *(3 marks)*

4. The potential at a point P which is a distance of 80 mm from a point charge X is +6.2 kV.
   a. Calculate: i the charge of X *(2 marks)*
   ii the potential at a distance of 1000 mm from X. *(2 marks)*
   b. Calculate the magnitude of the electric field strength at P. *(2 marks)*

22.6 Comparing electric fields and gravitational fields

Specification reference: 3.7.1

Table 1 summarises the similarities and differences between electric and gravitational fields.

▼ **Table 1** *Similarities and differences between gravitational and electric fields*

	Gravitational fields	Electrostatic fields
	Similarities	
line of force or a field line	path of a free test mass in the field	path of a free positive test charge in the field
inverse-square law of force	Newton's law of gravitation $F = \dfrac{Gm_1m_2}{r^2}$	Coulomb's law of force $F = \dfrac{Q_1Q_2}{4\pi\varepsilon_0 r^2}$
field strength	force per unit mass, $g = \dfrac{F}{m}$	force per unit + charge, $E = \dfrac{F}{q}$
potential energy of two point mass or charges	$E_P = \dfrac{-Gm_1m_2}{r}$	$E_P = \dfrac{Q_1Q_2}{4\pi\varepsilon_0 r}$
potential	work done per unit mass moved from infinity	work done per unit + charge moved from infinity
uniform fields	Field strength is the same everywhere, field lines parallel and equally spaced	
radial fields	due to a point mass or a uniform spherical mass M, $g = \dfrac{GM}{r^2}$ $V = \dfrac{-GM}{r}$	due to a point charge Q, $E = \dfrac{Q}{4\pi\varepsilon_0 r^2}$ $V = \dfrac{Q}{4\pi\varepsilon_0 r}$

> **Revision tip**
> A force field is a region in which a body experiences a non-contact force.

Summary questions

1. State two differences between a uniform field and a radial field. *(2 marks)*

2. The equations for the field strength and potential of a point charge or a point mass can be written as $Z = kYr^n$ where Z is the field strength or potential. Complete the table below by inserting the correct symbols for Y and k and the correct value of n. *(2 marks)*

	Z	Y	k	n
Gravitational field strength	g			
Electric field strength				
Gravitational potential	V			
Electric potential		Q		

3. a. Express the unit of $\dfrac{1}{4\pi\varepsilon_0}$ in SI base units. *(2 marks)*
   b. The escape velocity from a planet of radius R is $\sqrt{(2gR)}$. Write down an equivalent equation for the escape velocity of a charged particle from an oppositely charged sphere of radius R. *(1 mark)*

Chapter 22 Practice questions

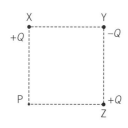

▲ Figure 1

1 Two point charges S and T at separation d exert a force of attraction F on each other. The charge on T is then doubled without changing the charge on S then their separation is doubled.

Which one of the following alternatives, **A–D**, gives the force of attraction after the changes?

A $\frac{1}{4}F$ **B** $\frac{1}{2}F$ **C** $2F$ **D** $4F$ (1 mark)

2 Three point charges X, Y, and Z are placed at the vertices of a square with X and Z at opposite vertices as shown in Figure 1. Their charges are $+Q$ for X and for Z and $-Q$ for Y.

Point P in Figure 1 is the vertex of the square where there is no charge. Which one of the following alternatives, **A–D**, gives the direction of the resultant electric field strength at P?

A Parallel to PY away from Y

B Parallel to PY towards Y

C At 45° to PX to the left of PX

D At 45° to PZ below PZ (1 mark)

3 Which one of the following alternatives, **A–D**, gives a correct unit for $G\varepsilon_0$?

A $kg\,C^{-2}$ **B** $kg^2\,C^{-2}$ **C** $C\,kg^{-2}$ **D** $C^2\,kg^{-2}$ (1 mark)

4 Figure 2 shows how the potential outside a positively charged metal sphere of radius 0.20 m varies with distance r from the centre of the sphere.

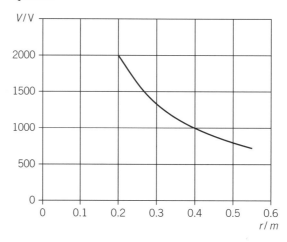

▲ Figure 2

a The potential at and outside the sphere is inversely proportional to r. Calculate the potential at a distance of 0.25 m from the centre of the sphere. (2 marks)

b The electric field at the surface and outside the sphere is the same as if the charge Q on the sphere was concentrated at its centre.

i Calculate the charge on the surface of the sphere. (2 marks)

ii Calculate the electric field strength at a point P which is a distance of 0.25 m from the centre of the sphere. (2 marks)

c An uncharged insulated metal rod XY of length 0.10 m is placed near the sphere along a radial line such that its nearest end X is at P.

i Explain why end X of the rod becomes negatively charged and end Y becomes positively charged. (4 marks)

40

Electric fields

▲ Figure 3

    ii Sketch a second curve on a copy of Figure 2 to show how the potential varies along the radial line from the surface of the sphere through X and through Y to a distance of 0.5 m from the centre of the sphere. *(4 marks)*

    iii Discuss the effect the rod has had on the electric field strength at and near P. *(2 marks)*

5 A charged polystyrene ball B is suspended on the end of an insulating vertical thread between two vertical parallel plates as shown in Figure 3.

When a potential difference of 2800 V is applied between the two plates, the ball is attracted towards one of the plates and the thread is displaced through an angle θ of 5.5°.

  a **i** Show that the force F on the ball due to the electric field is equal to $mg \tan \theta$, where m is the mass of the ball. *(2 marks)*

    ii The mass of the ball is 0.82 g and the perpendicular distance d between the plates is 35 mm. Calculate the charge on the ball. *(4 marks)*

  b When the distance d between the plates is increased, describe and explain the effect this change has on the ball. *(3 marks)*

6 Figure 4 shows two oppositely charged point charges A and B at a fixed separation of 0.25 m. Their total charge is +4.0 nC.

▲ **Figure 4**

  a The total electric potential is zero at a point P which is 0.10 m from A along the line AB.

Calculate the charge of A and the charge of B. *(4 marks)*

  b Copy Figure 4 and draw on your copy:

    i three lines of force, each one from one charge to the other, including one through P *(1 mark)*

    ii an equipotential line through P. *(2 marks)*

23.1 Capacitance
23.2 Energy stored in a charged capacitor
23.3 Charging and discharging a capacitor through a fixed resistor

Specification reference: 3.7.4.1; 3.7.4.3; 3.7.4.4

The **capacitance C of a capacitor is defined as the charge stored per unit pd.**

The unit of capacitance is the farad (F), which is equal to one coulomb per volt. Note that $1.0\,\mu F = 1.0 \times 10^{-6}\,F$.

For a capacitor that stores charge Q at pd V, its capacitance can be calculated using the equation

$$C = \frac{Q}{V}.$$

Charging a capacitor at constant current

In Figure 1, after the switch is closed, the variable resistor is continually adjusted to keep the microammeter reading constant and the capacitor pd is measured at regular intervals using a stopwatch and a high-resistance voltmeter.

The charge Q on the capacitor can be calculated using the equation $Q = It$, where I is the current and t is the time since the switch was closed.

A graph of charge stored, Q, on the y-axis against pd, V, on the x-axis gives a straight line passing through the origin. The gradient of the line is equal to the capacitance.

▲ **Figure 1** *Investigating capacitors*

> **Revision tip**
> Capacitance is measured in μF, nF, or pF. When you use $C = \frac{Q}{V}$ remember that charge in μC and pd in V gives capacitance in μF.

Energy stored in a charged capacitor

The energy stored by the capacitor when its pd is V and it stores charge Q is given by

$$E = \frac{1}{2}QV.$$

> **Maths skill**
>
> 1 Using $Q = CV$ or $V = \frac{Q}{C}$ you can write the above equation as $E = \frac{1}{2}CV^2$ or $E = \frac{1}{2}\frac{Q^2}{C}$.
>
> 2 Figure 2 shows how the pd increases as the charge stored increases from zero. The total work done is given by the total area of the triangle under the line which $= \frac{1}{2} \times$ base $\times$ height $= \frac{1}{2}QV$.
>
> 3 In the charging process, the battery transfers energy QV to the circuit. Thus, 50% of the energy supplied by the battery $= \frac{1}{2}QV$ is stored in the capacitor. The other 50% is wasted due to resistance in the circuit as it is transferred to the surroundings when the charge flows in the circuit.

▲ **Figure 2** *Energy stored in a capacitor*

Capacitor discharge through a fixed resistor

When a capacitor discharges through a fixed resistor, the current and the charge decrease exponentially with time.

For capacitor discharge through a resistor of fixed resistance R, when the charge is Q and the current is $I = \dfrac{dQ}{dt}$,

the capacitor pd $\dfrac{Q}{C}$ + the resistor pd $(IR) = 0$.

So $I = -\dfrac{Q}{RC}$. Therefore, $\dfrac{dQ}{dt} = -\dfrac{Q}{CR}$.

The mathematical solution is where $Q = Q_0 e^{-\frac{t}{RC}}$ where Q_0 is the initial charge, and e is the exponential function (sometimes written 'exp').

▲ **Figure 3** *Capacitor discharging*

> **Maths skill: Data analysis using natural logarithms**
>
> 1. The quantity RC is called the **time constant** for the circuit. At time $t = RC$ after the start of the discharge, the charge falls to $0.37 (= e^{-1})$ of its initial value. Note the half-life of the discharge $= RC \ln 2$.
> 2. Taking natural logs of both sides of the above equation gives $\ln Q = \ln Q_0 - \dfrac{t}{RC}$.
>   A graph of $\ln Q$ on the y-axis against t on the x-axis gives a straight line with a negative gradient $= -\dfrac{1}{RC}$.

> **Revision tip**
> An exponential decrease graph starts at an intercept with the vertical axis and is asymptotic with the time axis (i.e., the curve approaches but never cuts the time axis).

Charging a capacitor through a fixed resistor

1. At any instant during the charging process,
   the source pd V_0 = the resistor pd + the capacitor pd.
   Therefore, $V_0 = IR + \dfrac{Q}{C}$ at any instant.
2. The initial current $I_0 = \dfrac{V_0}{R}$, assuming that the capacitor is initially uncharged.
3. At time t after charging starts, $I = I_0 e^{-\frac{t}{RC}}$

   Rearranging this gives $Q = CV_0 \left(1 - e^{-\frac{t}{RC}}\right)$

   At time $t = 0$, $e^{-\frac{t}{RC}} = 1$, so $Q = 0$ and the capacitor pd $V = 0$.
   As time $t \to \infty$, $e^{-\frac{t}{RC}} \to 0$ so $Q \to Q_0$ and $V \to V_0$.

> **Worked example**
>
> A 470 µF capacitor is charged to a pd of 6.0 V then discharged through a 220 kΩ resistor using a circuit as shown in Figure 1. Calculate the pd after 200 s.
>
> When $t = 200$ s,
>
> $\dfrac{t}{RC} = \dfrac{200\,\text{s}}{220 \times 10^3 \times 470 \times 10^{-6}\,\text{F}}$
>
> $= 1.93$
>
> Therefore $V = V_0 e^{-\frac{t}{RC}} = 6.0\,\text{V} \times e^{-1.93}$
> $= 0.87\,\text{V}$

> **Summary questions**
>
> 1. A 47 µF capacitor is charged by using a constant current of 1.5 µA to a pd of 6.0 V. Calculate:
>   a. the charge stored on the capacitor at 6.0 V *(1 mark)*
>   b. the time taken. *(1 mark)*
>
> 2. A 22 µF capacitor was charged by connecting it to a 3.0 V battery. The battery was then disconnected from the charged capacitor and a 10 µF capacitor which was initially uncharged was then connected across the charged capacitor.
>   a. Calculate the charge stored on the 22 µF capacitor when it was connected to the battery. *(1 mark)*

Capacitors

▲ **Figure 4** *Capacitor charging*

b The pd across the capacitors after they were connected together was less than 3.0 V.
  i Explain why the pd was less than 3.0 V *(2 marks)*
  ii Calculate the pd across the capacitors after they were connected together. *(4 marks)*

3 An uncharged 10 μF capacitor is connected to a 6.0 V battery. Calculate:
  a i the charge and energy stored in the capacitor *(2 marks)*
  ii the energy supplied by the battery. *(1 mark)*
  b Explain the difference between the energy supplied by the battery and the energy stored in the capacitor. *(2 marks)*

4 A 47 μF capacitor is charged by connecting it to a 6.0 V battery then discharged through a 200 kΩ resistor.
  a i Calculate the time constant of the circuit. *(1 mark)*
  ii Estimate how long the capacitor would take to discharge to 2 V. *(2 marks)*
  b Calculate the pd and the discharge current 5.0 s after the discharge started. *(3 marks)*

5 An uncharged 68 μF capacitor is charged to a pd of 5.0 V through a 20 kΩ resistor. Calculate:
  a i the initial charging current *(1 mark)*
  ii the current when the pd across the capacitor is 3.0 V. *(2 marks)*
  b the time taken for the pd to increase from zero to 3.0 V. *(3 marks)*

23.4 Dielectrics
Specification reference: 3.7.4.2

Dielectrics are insulators used to increase the capacitance of a capacitor. When a dielectric is placed between two oppositely charged parallel plates connected to a battery, each molecule of the dielectric becomes **polarised** and forms **a dipole**. This means that its electrons are pulled slightly towards the positive plate as shown in Figure 1. So the surface of the dielectric facing the positive plate becomes negatively charged and the other surface of the dielectric becomes positively charged. As a result, the positive side of the dielectric attracts more electrons from the battery onto the negative plate. and the negative side of the dielectric pushes electrons back to the battery from the positive plate. So more charge is stored on the plates. In some dielectric substances, the molecules are already polarised. These molecules are called polar molecules. The electric field makes them turn so their electrons become nearer the positive plate. As a result, more charge is stored on the plates as explained above.

Relative permittivity
The **relative permittivity** (or dielectric constant) of a dielectric substance,

$$\varepsilon_r = \frac{\text{charge stored with dielectric present, } Q}{\text{charge stored without dielectric, } Q_o}$$

Capacitor design
For a parallel-plate capacitor with dielectric between the plates, its capacitance $C = \dfrac{A\varepsilon_0\varepsilon_r}{d}$ where A is the surface area of each plate and d is the spacing between the plates. A large capacitance can be achieved by increasing A, decreasing d and using a dielectric which has a relative permittivity as large as possible.

Dielectrics and alternating potential differences
In an alternating electric field, at low frequencies, the dipoles alternate in phase with the field regardless of the mass of the dipoles. As the frequency increases, the mass of the particles being moved by the field determines if they respond. The greater the mass of each particle, the greater its inertia is, and so the frequency at which it ceases to respond is lower.

▲ **Figure 1** *Dielectric action*

Summary questions
$\varepsilon_0 = 8.85 \times 10^{-12}\,\text{Fm}^{-1}$

1. A parallel-plate capacitor consists of two insulated metal plates separated by an air gap. A battery in series with a switch is connected to the plates. The capacitor is charged by closing the switch to charge the capacitor to a pd V.
   a. When the separation of the plates is increased, state and explain the change that takes place to:
      i. the capacitance C of the capacitor (2 marks)
      ii. the pd between the plates. (2 marks)
   b. State and explain the change that takes place to the energy stored by the capacitor when the plates are moved further apart. (2 marks)

2. Explain why a dielectric material in a capacitor increases its capacitance compared with the same capacitor with the same plate spacing and without dielectric. (4 marks)

Chapter 23 Practice questions

1. A capacitor of capacitance C is charged to a potential difference V then discharged through a variable resistor which is adjusted to keep the current at a constant value I for as long as possible. Which one of the following alternatives, **A–D**, gives the time taken for the discharge?

   A $\dfrac{V}{CI}$ **B** $\dfrac{CV}{I}$ **C** $\dfrac{I}{CV}$ **D** $\dfrac{IC}{V}$ *(1 mark)*

2. A uncharged capacitor is connected in series with a resistor, an open switch, and a battery. When the switch is closed, the capacitor charges to the same pd as the battery.
   Which one of the following alternatives, **A–D**, describes how the charge and the current changes with time? *(1 mark)*

	charge	current
A	Increases at a decreasing rate	Increases at a decreasing rate
B	Increases at an increasing rate	Increases at a decreasing rate
C	Increases at a decreasing rate	Decreases at a decreasing rate
D	Increases at an increasing rate	Decreases at an increasing rate

3. A capacitor charges through a resistor R from a battery to a potential difference V_B. The potential difference across the capacitor after time T from the start of the charging process is $0.40V_B$. Which one of the following alternatives, **A–D**, gives the potential difference across the capacitor after a further time T?

   A $0.50V_B$ **B** $0.60V_B$ **C** $0.64V_B$ **D** $0.80V_B$ *(1 mark)*

4. The circuit in Figure 1 is used to charge a capacitor using a constant current.

   ▲ **Figure 1**

   a. The current is maintained at 30 µA and the voltmeter reading is measured every 15 seconds until the capacitor is fully charged. The measurements are shown in the table below.

Time t / s	0	15	30	45	60	75	90
Capacitor pd V/V	0	0.51	1.02	1.50	1.98	2.52	3.01
Charge stored Q / mC	0	0.45					

   i Complete the last row of the table by calculating Q at time t. *(1 mark)*

   ii Plot a graph of Q on the y-axis against V on the x-axis. *(3 marks)*

   iii Use your graph to determine the capacitance of the capacitor. *(2 marks)*

   b. Discuss how the voltmeter reading would have changed with time if the variable resistor had not been adjusted. *(2 marks)*

Capacitors

5 A 2200 µF capacitor is charged to a potential difference of 5.5 V.

  a Calculate the energy stored in the capacitor. *(2 marks)*

  b i The charged capacitor is to be connected to a 5.0 V electric motor which has a thread attached to its axis supporting a small object of mass 3.8 g at its lower end.
Estimate the maximum possible height the object could be raised when the capacitor is discharged through the motor. *(3 marks)*

    ii Give two physical reasons why the height gain of the object would be less than your estimate above. *(2 marks)*

6 The potential difference across a capacitor was measured as it discharged through a 47 kΩ resistor. The measurements are shown in the table below:

Time t / s	0	30	60	90	120	150	180
Capacitor pd V/ V	5.95	4.55	3.60	2.70	2.15	1.70	1.35
ln V	1.78	1.52					

  a Use the equation $V = V_0 e^{-\frac{t}{RC}}$, where V_0 is the initial pd across the capacitor to show that a graph of ln V against time t should be straight line. *(3 marks)*

  b i Complete the last row of the table and use the results in the table to plot a graph of ln V against time t. *(5 marks)*

    ii Use your graph to determine the capacitance of the capacitor. *(4 marks)*

    iii Calculate the time constant of the circuit. *(1 mark)*

7 a A dielectric substance is inserted into the space between the plates of a parallel-plate capacitor which is connected to a battery.

    i Describe the effect on the dielectric of the electric field between the capacitor plates. *(3 marks)*

    ii Explain why the charge stored by the capacitor increases when the dielectric is inserted. *(3 marks)*

  b The battery is disconnected from the capacitor without altering the charge stored. When the dielectric is removed from the capacitor:

    i describe the effect on the capacitance of the capacitor *(1 mark)*

    ii state and explain the effect on the potential difference and the energy stored by the capacitor. *(5 marks)*

  c A 47 µF capacitor has an effective surface area of 1.2 m² and it contains a dielectric which has a relative permittivity of 7. Calculate the spacing between the capacitor plates, assuming the dielectric fills the entire space between the plates. *(2 marks)*

8 A 68 µF capacitor is charged to a pd of 6.0 V and then discharged through a 50 kΩ resistor.

  a Calculate the time taken after the discharge started for the pd to decrease to 1.5 V. *(3 marks)*

  b Estimate the resistance of the resistor that you would use in place of the 50 kΩ resistor if the discharge to 1.5 V is to take about 20 s. *(2 marks)*

24.1 Current-carrying conductors in a magnetic field

Specification reference: 3.7.5.1

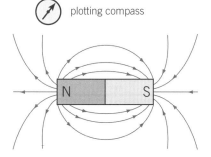

▲ **Figure 1** *The magnetic field near a bar magnet*

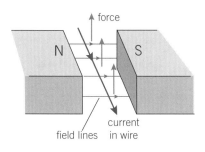

▲ **Figure 2** *The motor effect*

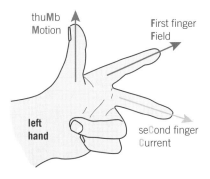

▲ **Figure 3** *Fleming's left-hand rule*

Revision tip
$F = BIl$ applies only when B and I are at right angles.

Magnetic field patterns

A magnetic field is a force field surrounding a magnet or current-carrying wire which acts on any other magnet or current-carrying wire placed in the field. The magnetic field of a bar magnet is strongest at its ends which are referred to as **north-seeking** and **south-seeking** 'poles'. A **line of force** (or magnetic field line) of a magnetic field is a line along which a north pole would move in the field.

The force on a current-carrying wire in a magnetic field

A current-carrying wire placed at a non-zero angle to the lines of force of an external magnetic field experiences a force due to the field. This effect is known as the **motor effect**. The force is perpendicular to the wire and to the lines of force.

The motor effect can be tested using the simple arrangement shown in Figure 3. The wire is placed between opposite poles of a U-shaped magnet so it is at right angles to the lines of force of the magnetic field. When a current flows, the section of the wire in the magnetic field experiences a force that pushes it out of the field. The force is:

- greatest when the wire is at right angles to the magnetic field
- zero when the wire is parallel to the magnetic field.

The direction of the force can be related to the direction of the field and to the direction of the current using **Fleming's left-hand rule** shown in Figure 3. If the current is reversed or if the magnetic field is reversed, the direction of the force is reversed.

The magnitude of the force on the wire is proportional to:

1 the current I
2 the length l of the wire.

The **magnetic flux density** B of the magnetic field, sometimes referred to as the strength of the magnetic field, is defined as the force per unit length per unit current on a current-carrying conductor at right angles to the magnetic field lines.

Therefore, for a wire of length l carrying a current I in a uniform magnetic field B at 90° to the field lines, the force F on the wire is given by

$$F = BIl$$

1 The unit of B is the tesla (T), equal to $1\,\text{Nm}^{-1}\text{A}^{-1}$.
2 The direction of the force is given by Fleming's left-hand rule. See Figure 3.

Applications
1 The electric motor

The simple electric motor consists of a coil of insulated wire which spins between the poles of a U-shaped magnet.

When a direct current passes round the coil:

- the wires at opposite edges of the coil are acted on by forces in opposite directions
- the force on each edge makes the coil spin about its axis.

Current is supplied to the coil via a split-ring commutator. The direction of the current round the coil is reversed by the split-ring commutator each time the coil rotates through half a turn. This ensures the current along an edge changes direction when it moves from one pole face to the other. As shown in Figure 4, the result is that the force on each edge continues to turn the coil in the same direction.

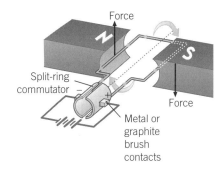

▲ **Figure 4** *In an electric motor*

2 The couple on a coil in a magnetic field

Consider a rectangular current-carrying coil of length l and width w in a uniform horizontal magnetic field, as shown in Figure 5. The coil has n turns of wire and can rotate about a vertical axis. When the current in the coil is I:

- Each long side therefore experiences a horizontal force $F = (BIl)n$ in opposite directions at right angles to the field lines.
- The pair of forces acting on the long sides form a couple. If the plane of the coil is at angle α to the field lines, the torque of the couple = $Fw\cos\alpha$ = $BIlnw\cos\alpha$ = $BIAn\cos\alpha$, where the coil area $A = lw$.

If $\alpha = 0$ (i.e., the coil is parallel to the field), the torque = $BIAn$ as $\cos 0 = 1$.

If $\alpha = 90°$ (i.e., the coil is perpendicular to the field), the torque = 0 as $\cos 90° = 0$.

a *top view of coil parallel to field*

b *top view of coil at angle α to field*

▲ **Figure 5** *Couple on a coil*

Summary questions

1. A straight vertical wire of length 80 mm carrying a constant current of 4.3 A downwards is in a uniform horizontal magnetic field which acts due North. The wire experiences a force of 29 mN.
   a Determine the magnitude of the magnetic flux density of the magnetic field *(2 marks)*
   b State the direction of the force. *(1 mark)*

2. A square coil of length 60 mm has 50 turns. The coil is in a uniform horizontal magnetic field of flux density 55 mT which is perpendicular to the plane of the coil. A current of 5.9 A is in the coil. Sketch the arrangement and determine the magnitude and direction of the force on each side of the coil. *(4 marks)*

3. A horizontal cable suspended between two pylons is aligned in a north-south plane. The Earth's magnetic field at this position has a vertical component of 66 µT and is in a downwards direction. Determine the magnitude of the force per unit length on the cable when the current in it is 65 A. *(2 marks)*

24.2 Moving charges in a magnetic field
24.3 Charged particles in circular orbits

Specification reference: 3.7.5.1; 3.7.5.2

> **Revision tip**
> The equation $F = BQv\sin\theta$ (when the direction of motion of a charged particle in a magnetic field is at angle θ to the lines of the field) is not required for this specification.

> **Revision tip**
> The direction of the magnetic force on a moving electron is given by Fleming's left-hand rule. For negative particles, remember that the current direction is opposite to the direction in which the particles move.

Force on a moving charge in a magnetic field

For a particle of charge Q moving through a uniform magnetic field at speed v in a perpendicular direction to the field, the force on the particle is given by

$$F = BQv$$

The velocity selector

This apparatus has a magnetic field and an electric field perpendicular to each other. A charged particle entering the fields perpendicular to both fields is undeflected if its speed is such that the magnetic field force on it equals the electric field force on it.

$$B_s Q v = \frac{QV_P}{d}$$

In a velocity selector, ions moving at different speeds in a narrow beam are directed into the arrangement. The ions moving at speed $v = \frac{V_P}{B_s d}$ (from the above equation) pass through undeflected and can therefore pass through a collimator slit that stops all the other ions in the beam.

▲ **Figure 1** Electrons in a magnetic field

Charged particles in circular orbit

The magnetic force on a charged particle moving in a magnetic field is always perpendicular to the velocity at any point along the path. The particle therefore moves in a circular path with the force always acting towards the centre of curvature of the circular path.

Figure 2 shows a beam of electrons in a uniform magnetic field on a circular path. At any point on the path, the magnetic force BQv = the centripetal force $\frac{mv^2}{r}$ where r is the radius of the circular path.

Rearranging this equation gives $r = \frac{mv}{BQ}$.

▲ **Figure 2** A circular orbit in a magnetic field

Applications

The **cyclotron** consists of two hollow D-shaped electrodes (referred to as 'dees') in a vacuum chamber. A uniform magnetic field and an alternating pd are applied to the dees as shown in Figure 3.

Charged particles entering one of the dees near the centre of the cyclotron are forced on a semi-circle by the magnetic field, causing them to emerge from the dee. As they cross into the other dee, the alternating voltage reverses so they are accelerated into the other dee where they are once again forced by the magnetic field on a semi-circle with a larger radius. On emerging from this dee, the process is repeated until the particles leave one of the dees through a gap at the edge.

The particles are accelerated each time they cross from one dee to the other because the time taken t to complete each semi-circle equals $\frac{\pi r}{v}$ where v is their speed and r is the radius of the semi-circle.

The radius $r = \frac{mv}{BQ}$ therefore $t = \frac{m\pi}{BQ}$ which is independent of the particle's speed and is equal to the time taken for half a cycle of the alternating voltage.

The mass spectrometer

In a mass spectrometer, the atoms of the sample are ionised and directed in a narrow beam at the same velocity (after passing through a velocity selector) into a uniform magnetic field. Each ion is deflected in a semi-circle by the magnetic field onto a detector. The radius of curvature of the path of each ion depends on the specific charge $\frac{Q}{m}$ of the ion in accordance with the equation $r = \frac{mv}{BQ}$. Each type of ion is deflected by a different amount onto the detector. The detector is linked to a computer which is programmed to show the relative abundance of each type of ion in the sample.

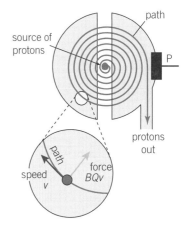

▲ Figure 3 *The cyclotron*

Summary questions

$e = 1.6 \times 10^{-19}$ C

1. Calculate the force on an electron that enters a uniform magnetic field of flux density 150 mT at a velocity of 8.0×10^6 ms^{-1} at an angle of:
   a 90° *(1 mark)*
   b 0° to the field. *(1 mark)*

2. A beam of protons at a speed of 3.2×10^7 ms^{-1} is directed into a uniform magnetic field of flux density 85 mT in a direction perpendicular to the field lines. The protons move on a circular path in the field.
   mass of proton = 1.67×10^{-27} kg
   a Calculate the radius of curvature of the path. *(3 marks)*
   b The flux density is halved and the kinetic energy of the protons is halved. Determine the effect of this change on the radius of curvature of the path. *(3 marks)*

3. In a mass spectrometer, a beam of ions was directed in a velocity selector and the ions that emerged had a speed of 9.5×10^5 ms^{-1}.
   a The electric field strength in the velocity selector was 82 kVm^{-1}. Calculate the magnetic flux density in the velocity selector. *(2 marks)*
   b These ions were then directed into a uniform magnetic field of flux density 910 mT where they were deflected in a semi-circular path of diameter 370 mm on to the detector. Calculate the specific charge of these ions. *(3 marks)*

Chapter 24 Practice questions

1. A velocity selector is an arrangement in which a uniform electric field is perpendicular to a uniform magnetic field. The magnetic flux density is B and the speed of the charged particles that pass through undeflected is v. Which one of the following alternatives, **A–D**, gives the electric field strength in this situation?

   A vB **B** $\dfrac{B}{v}$ **C** $\dfrac{v}{B}$ **D** B *(1 mark)*

2. In an electron tube, electrons of mass m are accelerated from rest through a potential difference V before they enter a uniform magnetic field in a direction perpendicular to the field lines. The electrons move on a circular path of radius r in the field when the magnetic flux density is B. Which one of the following alternatives, **A–B**, gives the equation for r in terms of V and B?

   A $\sqrt{\dfrac{2mV}{eB}}$

   B $\sqrt{\dfrac{2mV}{eB^2}}$

   C $\sqrt{\dfrac{2meV}{B^2}}$

   D $\sqrt{\dfrac{2meV}{eB}}$ *(1 mark)*

3. In a high-energy collision in a magnetic field, several particles and antiparticles emerge from the collision, each on a different curved track. One of the tracks P is created by a proton. A second track X curves in the same direction as P with a smaller radius of curvature than P. Which one of the following statements about the particle that created track X is correct?

   A X could be a proton with a greater momentum than P

   B X could be an antiproton with less momentum that P

   C X could be a π^+ meson with less momentum than P

   D X could be a π^- meson with greater momentum than P *(1 mark)*

4. **a** A charged particle in a uniform magnetic field is moving along a circular path at a constant speed. Explain why the particle moves a constant speed even though it is acted on by a force due to the magnetic field. *(3 marks)*

   b The most abundant isotope of neon is $^{20}_{10}$Ne. A sample of neon gas is analysed in the spectrometer. Each neon ion carries a single charge of $+1.6 \times 10^{-19}$ C. The neon ions enter the magnetic field of the spectrometer at a velocity of 7.3×10^5 m s^{-1} and they are deflected by the magnetic field through 180° onto a detector.

   i The ions strike the detector 0.31 m from the point **P** at which the ion beam enters the magnetic field. Calculate the magnetic flux density of the magnetic field. *(3 marks)*

   ii Neon ions are also detected at a distance of 0.34 m from point **P**. Calculate the relative atomic mass of these neon ions. *(2 marks)*

   iii Explain why these ions have a different mass from the $^{20}_{10}$Ne ions. *(2 marks)*

5 In an earphone, a circular coil of insulated wire is in a radial magnetic field as shown in Figure 1. The coil is supported on a thin flexible disc attached to the ring-shaped pole.

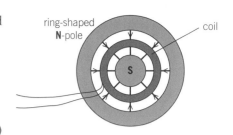
▲ Figure 1

   a The coil has 60 turns and a diameter of 8.2 mm. The magnetic field between the poles has a magnetic flux density of 120 mT. Calculate the force on the coil due to the magnetic field when the current in the coil is 0.19 A. *(2 marks)*

   b i Explain why the coil vibrates when there is an alternating current in the coil. *(2 marks)*

     ii Discuss how the amplitude of the vibrations of the coil varies with the rms current in the coil. *(2 marks)*

6 A current balance consists of a rectangular frame ABCD balanced on pivots midway along AB and CD as shown in Figure 2. A small piece of wire W of mass 0.67 g is positioned along AB to balance the frame horizontally. End AD is positioned at the end of the solenoid at right angles to the solenoid axis.

▲ **Figure 2**

When a constant current of 4.5 A is passed through the solenoid and along end AD of the solenoid in series with one another, W is moved 32 mm towards the pivot to regain the balance. Given AB = CD = 240 mm and AD = BC = 95 mm, calculate:

   a the force on end AD due to the magnetic field of the solenoid *(4 marks)*

   b the magnetic flux density at the end of the solenoid. *(2 marks)*

7 In a direct-current electric motor, the coil has 160 turns wound on a rectangular frame of length 35 mm and width 28 mm. The magnetic field in which the coil spins has a magnetic flux density of 110 mT.

   a The current in the coil is 0.67 A. Calculate:

     i the force on each long side of the coil when the current in the coil is 0.67 A *(1 mark)*

     ii the maximum torque on the coil due to the magnetic field. *(2 marks)*

   b A pulley of diameter 5.4 mm fixed to the motor axle is used to raise a 1.4 N weight W attached to a thread wrapped round the pulley. Estimate the least current in the coil that would be necessary to raise the weight. *(4 marks)*

25.1 Generating electricity

Specification reference: 3.7.5.4

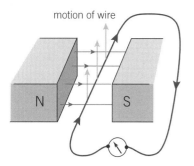

▲ Figure 1 *Generating an electric current*

Revision tip
An emf is induced whenever there is relative motion between the coil and the magnet. The wire must cut across the lines of the magnetic field for an emf to be induced in the wire.

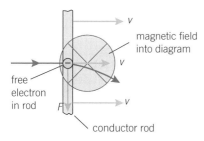

▲ Figure 2 *Deflection of electrons in a magnetic field*

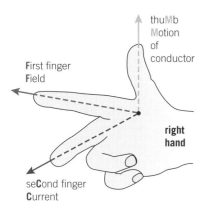

▲ Figure 3 *Fleming's right-hand rule*

Investigating electromagnetic induction

When a magnet is moved near a wire connected to a sensitive meter, an electromotive force (emf) is **induced** in the wire and a current passes through the meter. See Figure 1. The **induced emf** forces electrons round the circuit. The induced emf can be increased by:

- moving the wire faster
- using a stronger magnet
- making the wire into a coil and pushing the magnet in or out of the coil.

Energy changes

The rate of transfer of energy from the source of emf to the other components of the circuit is equal to the product of the induced emf and the current. This is because the induced emf × the current = energy transferred per unit charge from the source × the charge flow per second = energy transferred per second from the source.

The dynamo rule

Look at Figure 2. The magnetic field is into the plane of the diagram and the motion of the conductor is to the right. The electrons in the rod experience a force downwards due to the magnetic field so the top end becomes positive and the lower end becomes negative so an emf is induced in the rod. If the rod is part of a complete circuit, an induced current passes round the circuit. The direction of the induced current in the rod is given by **Fleming's right-hand rule**, as shown in Figure 3. The direction of the induced current is, in accordance with the current convention, opposite to the direction of the flow of electrons in the conductor.

Summary questions

1. A straight section XY of insulated wire is connected to a sensitive meter.
   a. Explain why the meter shows a brief reading when XY is moved between the poles of a U-shaped magnet. *(2 marks)*
   b. State one way in which the meter reading could be made larger. *(1 mark)*

2. In **Q1**, XY is vertical with X above Y. The magnetic field lines are horizontal in a direction from east to west. State and explain the direction of the induced current in the wire when it is moved in a direction from south to north without changing its vertical alignment. *(2 marks)*

3. A flat circular coil of wire is connected to a sensitive centre-reading meter. The coil is placed between the poles of a U-shaped magnet with its plane perpendicular to the field lines. The coil is then removed from between the poles then returned to its original position between the poles. Describe and explain how the meter reading changes if it is removed quickly and returned slowly. *(4 marks)*

25.2 The laws of electromagnetic induction

Specification reference: 3.7.5.3; 3.7.5.4

The solenoid rule

Figure 1 shows the magnetic field pattern of a current-carrying solenoid. If each end in turn is viewed from outside the solenoid:

- current passes a**N**ticlockwise (or cou**N**terclockwise) round the '**N**orth pole' end
- current passes clockwise round the 'south pole' end.

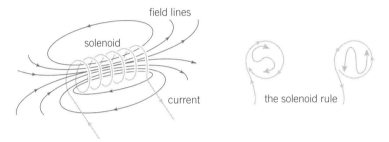

▲ **Figure 1** *The magnetic field near a solenoid*

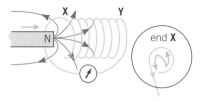

▲ **Figure 2** *Lenz's law*

Lenz's law

Consider the north pole of a bar magnet approaching end X of a coil, as shown in Figure 2. The induced current in the circuit creates a magnetic field in the coil. The induced polarity of end X must be a N-pole so as to repel the incoming N-pole. **Lenz's law states that the direction of the induced current is always such as to oppose the change that causes the current.**

Revision tip
The induced current could never be in a direction to help the change that causes it — that would mean producing electrical energy from nowhere, which is forbidden by conservation of energy.

Magnetic flux and flux linkage

Consider a uniform magnetic field of magnetic flux density B which is perpendicular to an area A. The **magnetic flux ϕ through this area $A = BA$.** The unit of magnetic flux is the **weber** (Wb), equal to $1\,\text{Tm}^2$. The **magnetic flux linkage** through a coil of N turns and area $A = N\phi = NBA$ When the magnetic field is at angle θ to the normal at the coil face, the flux linkage through the coil $= BAN\cos\theta$

Faraday's law of electromagnetic induction

Faraday's law of electromagnetic induction states that the induced emf in a circuit is equal to the rate of change of flux linkage through the circuit.

$$\text{Induced emf } \varepsilon = -N\frac{\Delta\phi}{\Delta t}$$

where $N\dfrac{\Delta\phi}{\Delta t}$ is the change of flux linkage per second. The minus sign represents the fact that the induced emf acts in such a direction as to oppose the change that causes it (as per Lenz's law).

The unit of flux change per second, the weber per second, is the same as the volt. Therefore, the weber is equal to 1 volt second. When a wire of length L moving at speed v cuts across the lines of a magnetic field of flux density B, the induced emf $= BLv$ (because the area swept out by the rod $= Lv$ so the flux swept out per second $= BLv$).

Summary questions

1. A horizontal metal rod of length 0.063 m is moved at a speed of 12 mms^{-1} across a vertical magnetic field of flux density 72 mT in a direction perpendicular to its length. Calculate:
   a. the magnetic flux swept out by the rod each second *(3 marks)*
   b. the induced emf in the rod as it cuts across the field lines. *(1 mark)*

2. A solenoid is connected to a centre-reading meter. When a bar magnet is brought near end X of a solenoid, the meter deflects briefly.
   a. Explain why the meter deflects briefly. *(3 marks)*
   b. State and explain how the meter reading changes when the magnet is moved away from end X of the solenoid. *(3 marks)*

25.3 The alternating current generator
25.4 Alternating current and power

Specification reference: 3.7.5.4; 3.7.5.5

▲ Figure 1 *The ac generator*

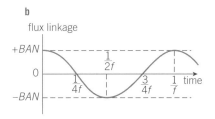

▲ Figure 2 *Flux linkage in a spinning coil*

Revision tip
The induced emf is zero when the sides of the coil move parallel to the field lines and maximum when the sides of the coil cut the field lines at right angles.

Revision tip
The peak-to-peak value of an alternating current or pd (from the top to the bottom of the wave) = 2 × the peak value. An oscilloscope may be used to measure the peak-to peak value of an alternating pd. See Topic 4.7, Using an oscilloscope.

The alternating current generator

A simple ac generator is shown in Figure 1. When the coil spins at a steady rate, the flux linkage changes continuously. At an instant when the normal to the plane of the coil is at angle θ to the field lines, **the flux linkage through the coil equals $BAN\cos\theta$** where B is the magnetic flux density, A is the coil area, and N is the number of turns on the coil. For a coil spinning at a steady frequency, f, $\theta = 2\pi ft$ at time t after $\theta = 0$. See Figure 2. The flux linkage = $BAN\cos 2\pi ft$. The induced emf $\varepsilon = \varepsilon_o \sin 2\pi ft = \varepsilon_o \sin \omega t$ where $\omega = 2\pi f$ and the peak emf, $\varepsilon_o = BAN\omega$. The gradient of the graph is the change of flux linkage per second $\dfrac{N\Delta\phi}{\Delta t}$, so is equal to $-\varepsilon$.

Back emf

A back emf, ε, is induced in the spinning coil of an electric motor because the flux linkage through the coil changes. The induced emf acts against the pd V applied to the motor in accordance with Lenz's law. At any instant, $V - \varepsilon = IR$, where I is the current through the motor coil and R is the circuit resistance.

Because the induced emf is proportional to the speed of rotation of the motor at low speed, the current is high because the induced emf is small, and at high speed, the current is low because the induced emf is high.

Alternating current and power

The **peak value** of an alternating current I_o (or pd V_o) is the maximum current (or pd) in either direction. Power (= I^2R) supplied to a resistor by an alternating current varies with time. At peak current I_o, maximum power is supplied equal to I_o^2R. The mean power = $\dfrac{1}{2}I_o^2R$.

The root mean square value of an alternating current I_{rms} is the value of direct current that would give the same heating effect as the alternating current in the same resistor.

Therefore $I_{rms}^2 R = \dfrac{1}{2} I_o^2 R$ so $I_{rms} = \dfrac{1}{\sqrt{2}} I_o$ and $V_{rms} = \dfrac{1}{\sqrt{2}} V_o$

Summary questions

1. The coil of an ac generator has 120 turns, a length of 22 mm, and a width of 18 mm. It spins in a uniform magnetic field of flux density 110 mT at a constant frequency of 50 Hz.
   a Calculate the maximum flux linkage through the coil. *(2 marks)*
   b Show that the peak voltage is 1.6 V. *(2 marks)*

2. An electric motor is used to raise a weight. The motor is connected in series with a battery and an ammeter. Explain why the motor current increases when the load on it is increased. *(3 marks)*

3. A mains heater gives a mean output power of 1.0 kW when the rms pd across it is 230 V. Calculate:
   a the rms current *(1 mark)*
   b the peak current *(1 mark)*
   c the peak power. *(2 marks)*

25.5 Transformers

Specification reference: 3.7.5.6

A transformer consists of two coils (the primary coil and the secondary coil.) wound on the same iron core. The symbol for the transformer is shown in Figure 1.

a *practical arrangement* b *transformer symbol*

▲ **Figure 1** *The transformer*

The transformer rule

$$\frac{V_s}{V_p} = \frac{N_s}{N_p}$$

where V_s is the secondary voltage, V_p is the primary voltage, N_s is the number of turns on the secondary coil, and N_p is the number of turns on the primary coil.

- A **step-up transformer** has more turns on the secondary coil than on the primary coil. So the secondary voltage is greater than the primary voltage.
- A **step-down transformer** has fewer turns on the secondary coil than on the primary coil. So the secondary voltage is less than the primary voltage.

Transformer efficiency

Transformers are almost 100% efficient because they have:

1. **low-resistance windings** to reduce power wasted due to the heating effect of the current
2. **a laminated core** (layers of iron separated by layers of insulator). Eddy currents induced in the core by the changing magnetic field in the core are reduced in this way, so their heating effect is reduced
3. **a core of 'soft iron'** which is easily magnetised and demagnetised. This reduces power wasted through repeated magnetisation and demagnetisation of the core.

The efficiency of a transformer = $\frac{\text{power delivered by the secondary coil}}{\text{power supplied to the primary coil}}$

$$= \frac{I_s V_s}{I_p V_p} \times 100\%$$

The efficiency of most transformers is almost 100%, therefore $I_s V_s = I_p V_p$ assuming the transformer is 100% efficient.

So the current ratio $\frac{I_s}{I_p} = \frac{V_p}{V_s} = \frac{N_p}{N_s}$.

Worked example

A transformer is used to step down 230 V mains to 12 V. When a 12 V, 36 W lamp is connected to the transformer's secondary coil, the lamp lights normally. When the lamp is on, the primary current in the transformer is 0.16 A. Calculate the efficiency of the transformer.

Efficiency $= \frac{I_s V_s}{I_p V_p} = \frac{36\,W}{230\,V \times 0.16\,A} = 0.98 \,(= 98\%)$

The grid system

Transmission of electrical power is more efficient at high voltage than at low voltage. This is because the current needed to deliver a certain amount of power is reduced if the voltage is increased. So the power wasted due to the heating effect of the current through the cables is reduced.

Revision tip
In a step-up transformer, the voltage is stepped up and the current is stepped down. In a step-down transformer, the voltage is stepped down and the current is stepped up.

Summary questions

1. In terms of electrical power, explain why the current through the primary coil of a transformer decreases when a device connected to the secondary coil is switched off. *(2 marks)*

2. **a** State three reasons why transformers are not 100% efficient. *(3 marks)*
   b Explain why power is transferred through the grid system more efficiently the higher the grid voltage is. *(5 marks)*

3. A transformer is designed to step down a primary voltage of 230 V to 6.0 V. Its primary coil has 2400 turns.
   a Calculate the number of turns in the secondary coil. *(2 marks)*
   b A 6.0 V, 24 W lamp is connected to its secondary coil. Calculate the current in the primary coil. *(3 marks)*

Chapter 25 Practice questions

1. A rectangular coil which has n turns and an area A is in a uniform magnetic field which is perpendicular to the plane of the coil. When the coil is removed from the field in a time t, a potential difference V is induced in the coil. Which one of the following alternatives, **A–D**, gives a correct expression for the magnetic flux density of the magnetic field?

   A $\dfrac{VAt}{n}$ **B** $\dfrac{At}{Vn}$ **C** $\dfrac{nA}{Vt}$ **D** $\dfrac{Vt}{nA}$ *(1 mark)*

2. Which one of the following combination of units has the same SI base units as the unit of magnetic flux?

   A T A^{-1} **B** T m^{-2} **C** V s   **D** V A^{-1} *(1 mark)*

3. A metal rod of length L moving at a constant velocity v cuts across a uniform magnetic field which has a magnetic flux density B. Which one of the lines **A–D** in the graph in Figure 1 shows how the induced emf V varies with time t during the time the rod is in the field?   *(1 mark)*

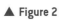

▲ Figure 1

4. A circular flat coil of radius 25 mm with 75 turns is connected to a data recorder. The coil is placed between the poles of a U-shaped magnet with the plane of the coil parallel to the pole faces. The coil is then rapidly removed from the field and the data recorder is used to measure the induced emf at 10 ms intervals. Figure 1 shows a graph of the measurements.

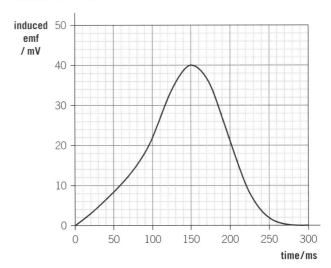

▲ Figure 2

   a. Explain why the area under the line in Figure 2 represents the change of flux linkage through the coil.   *(2 marks)*

   b. From the graph, estimate the change of flux linkage when the coil was removed from the magnetic field.   *(3 marks)*

   c. i. Calculate the magnetic flux density between the poles of the magnet.   *(3 marks)*

      ii. The diameter of the coil was estimated to within ± 0.5 mm. Estimate the uncertainty in the magnetic flux density calculated in part **c i**.   *(3 marks)*

5. An alternating current generator has a rectangular coil of area 2.3×10^{-3} m^2 with 180 turns. The coil rotates at a constant frequency of 50 Hz in a uniform magnetic field of flux density 92 mT with the coil axis perpendicular to the field lines.

   a. i. Calculate the maximum flux linkage in the coil.   *(2 marks)*

      ii. Calculate the peak emf S induced in the coil.   *(2 marks)*

b Sketch a graph to show how the induced emf varies with time over two cycles from when the flux linkage in the coil is a maximum. *(3 marks)*

c Describe how the graph would differ if the frequency had been increased to 90 Hz. *(2 marks)*

6 A transformer consists of a primary coil with 1800 turns and a secondary coil with 94 turns. A 12 V, 50 W lamp is connected to the secondary coil.

a i Calculate the rms potential difference of the alternating pd that must be applied to the primary coil to operate the lamp at its normal brightness. *(2 marks)*

ii Calculate the peak value of the pd applied to the primary coil. *(1 mark)*

b When the lamp is at normal brightness, the primary current has an rms value of 0.23 A. Calculate the efficiency of the transformer. *(2 marks)*

c State and explain two possible reasons why the transformer is not 100% efficient. *(2 marks)*

7 A speedometer in a vehicle consists of a metal disc that rotates between the poles of a U-shaped magnet when the vehicle is moving. A voltmeter is connected between the rim and the centre of the disc as shown in Figure 3. The frequency of rotation of the disc is proportional to the speed of the vehicle.

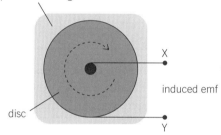

▲ Figure 3

a i Explain why an emf is induced between the rim and the centre of the disc when the disc is rotating. *(3 marks)*

ii Determine the polarity of the induced emf at X when the disc rotates clockwise. *(1 mark)*

b When the disc rotates at constant frequency, each radial element of the disc sweeps out an area equal to the area A of the disc in each complete rotation. Show that the induced emf is equal to BAf where B is the magnetic flux density of the magnetic field. *(3 marks)*

c The radius of the disc is 15 mm and the magnetic flux density is 270 mT. Calculate the induced emf when the frequency of rotation of the disc is 46 Hz. *(2 marks)*

8 A square coil of length 18 mm has 55 turns. The coil is placed in a uniform magnetic field of flux density 0.13 T with its plane perpendicular to the field lines. The coil was reversed in a time of 92 ms.

Calculate the magnitude of the induced emf. *(5 marks)*

26.1 The discovery of the nucleus

Specification reference: 3.8.1.1

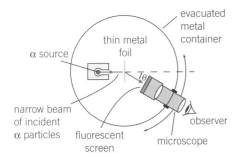

▲ **Figure 1** *Rutherford's α-scattering apparatus*

Figure 1 shows an outline of the arrangement Ernest Rutherford used in 1911 to probe the structure of the atom.

A thin metal foil in the path of a narrow beam of α particles scattered the particles. When the particles hit the atoms of a fluorescent screen in front of a microscope, each impact created a pinpoint of light. The number of α particles reaching the screen per minute was measured for different angles of deflection from zero to almost 180°. The measurements showed that:

1. most α particles passed straight through the foil with little or no deflection – about 1 in 2000 were deflected
2. about 1 in 10 000 particles were deflected through angles of more than 90°.

Rutherford concluded that:

- Most of the atom's mass is concentrated in a small region, the nucleus, at the centre of the atom.
- The nucleus is positively charged because it repels α particles (which carry positive charge) that approach it too closely.
- By testing foils of different metal elements, the magnitude of the charge of a nucleus is $+Ze$, where e is the charge of the electron and Z is the atomic number of the element.

Estimate of nuclear diameter

In a head-on impact, the α particle stops momentarily at the least distance of approach, d. At this point, the potential energy of the α particle in the electric field of the nucleus is equal to the initial kinetic energy, E_K, of the α particle. Therefore, $E_K = \dfrac{Q_\alpha Q_N}{4\pi\varepsilon_0 d}$, where Q_α is the charge of the α particle ($+2e$) and Q_N is the charge of the nucleus ($= +Ze$). For an α particle with kinetic energy of 8×10^{-13} J directed at an aluminium nucleus ($Z = 13$), using this formula gives a least distance of approach of about 7×10^{-15} m.

> **Synoptic link**
>
> A more accurate method of determining the radius of a nucleus is explained in Topic 26.1, The discovery of the nucleus.

> **Revision tip**
>
> Nuclear diameter is of the order of 10^{-15} m.

> **Summary questions**
>
> 1. **a** In the Rutherford α particle scattering experiment, only a small fraction of the α particles were deflected through large angles. What did Rutherford deduce about the atom from this discovery? *(3 marks)*
>
>   **b** Discuss how the results would have differed if the α particles in the beam had a wide range of speeds. *(3 marks)*
>
> 2. An α particle collided with a nucleus and was deflected by 90°.
>
>   **a** Sketch the path of the α particle as it approaches the nucleus and then moves away from it. Label the nucleus N and label the point P on the path at which the α particle is nearest to the nucleus. *(3 marks)*
>
>   **b** Describe how the kinetic energy and the potential energy of the α particle change as the particle approaches the nucleus then moves away from it. *(3 marks)*
>
> 3. Calculate the least distance of approach of an α particle to an aluminium nucleus ($Z = 13$) if the initial kinetic energy of the α particle is 5.1 MeV. *(3 marks)*
>
> $\varepsilon_0 = 8.85 \times 10^{-12}$ Fm^{-1}, $e = 1.6 \times 10^{-19}$ C

26.2 The properties of α, β, and γ radiation
26.3 More about α, β, and γ radiation
Specification reference: 3.8.1.2

Summary of the properties of α, β, and γ radiation

▼ **Table 1** *Nature and properties of α, β, and γ radiation*

	α radiation	β radiation	γ radiation
Nature	2 protons + 2 neutrons	β⁻ = electron (β⁺ = positron)	photon of energy of the order of MeV
Range in air	fixed range, depends on energy, up to ≈100 mm	range up to about 1 m	follows the inverse-square law
Deflection in a magnetic field	deflected	opposite direction to α particles; more easily deflected	not deflected
Absorption	stopped by paper or thin metal foil	stopped by approx 5 mm of aluminium	stopped or significantly reduced by several cm of lead
Ionisation	≈ 10^4 ions per mm in air at standard pressure	≈100 ions per mm in air at standard pressure	very weak ionising effect
Energy of each particle/photon	constant for a given source	varies up to a maximum for a given source	constant for a given source
Equations	$^A_ZX \rightarrow\, ^4_2\alpha + ^{A-4}_{Z-2}Y$	$^A_ZX \rightarrow\, ^0_{-1}\beta + ^A_{Z+1}Y + \bar{\nu}_e$	no change to A and Z

β⁺ emission

A positron (i.e., a β⁺ particle) is emitted from a proton-rich nucleus when a proton in the nucleus changes into a neutron. An electron neutrino v_e is emitted at the same time as the β⁺ particle is created.

$$^A_ZX \rightarrow\, ^0_{+1}\beta + ^A_{Z-1}Y + v_e$$

Electron capture

Some proton-rich nuclides can capture an inner-shell electron. In this event, a proton in the nucleus changes to a neutron and an electron neutrino is emitted. The inner-shell vacancy is filled by an outer-shell electron with the emission of an X-ray photon.

$$^A_ZX + ^0_{-1}e \rightarrow\, +^A_{Z-1}Y + v_e$$

The inverse-square law for γ radiation

The **intensity** I of the radiation is the radiation energy per second passing normally through unit area.

- For a point source that emits n γ photons per second, each of energy hf, the radiation energy per second from the source = nhf.
- At distance r from the source, the intensity $I = \dfrac{\text{radiation energy per second}}{\text{total area}}$ = $\dfrac{k}{r^2}$, where $k = \dfrac{nhf}{4\pi}$.

Revision tip
See Topic 26.3, in the Student Book for the experimental verification of the inverse square law for gamma radiation.

Summary questions

1 **a** A certain isotope emits β⁺ radiation. Describe in terms of protons and neutrons the changes that take place when a nucleus of this isotope emits a β⁺ particle. *(2 marks)*

   b A point source of γ radiation is placed 75 mm from the end of a Geiger tube. The corrected count rate was measured at 25.2 counts per second. Calculate the corrected count rate if the source was moved to a distance of 160 mm from the tube. *(3 marks)*

2 The astatine isotope $^{210}_{85}$At decays by electron capture to form an unstable isotope of polonium (Po) which then decays by emitting an α particle to form a stable isotope of lead (Pb). Write down the equation for:

   a the decay of $^{210}_{85}$At *(2 marks)*

   b the decay of the polonium nucleus. *(1 mark)*

26.4 The dangers of radioactivity

Specification reference: 3.8.1.2

The hazards of ionising radiation

Ionising radiation includes X-rays, protons, and neutrons as well as α, β, and γ radiation. High doses of ionising radiation kill living cells. Ionising radiation affects living cells because:

- it can destroy cell membranes which causes cells to die
- it can damage vital molecules such as DNA directly or indirectly by creating 'free radical' ions which react with vital molecules. Damaged DNA may cause cells to divide and grow uncontrollably, causing a tumour which may be cancerous. Damaged DNA in a sex cell (i.e., an egg or a sperm) may cause a mutation which may be passed on to future generations.

The biological effect of ionising radiation depends on the dose received and the type of radiation. The dose is measured in terms of the energy absorbed per unit mass of matter from the radiation. The same dose of different types of ionising radiation has different effects. For example, α radiation produces far more ions per millimetre than γ radiation in the same substance so it is far more damaging. However, α radiation from a source outside the body cannot penetrate the skin's outer layer of dead cells so is much less damaging than if the source were inside the body.

Background radiation

Background radiation occurs naturally due to various sources such as cosmic radiation and from radioactive materials in rocks, soil, and in the air. See Figure 1.

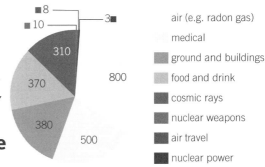

▲ Figure 1 Sources of background radiation in the UK in terms of the average radiation dose per person per year in μSv

Safe use of radioactive materials

Because radioactive materials produce ionising radiation, they must be stored and used with care. In addition, disposal of a radioactive substance must be carried out in accordance with specific regulations.

1. **Storage of radioactive materials** should be in lead-lined containers. The lead lining of a container must be thick enough to reduce the γ radiation from the sources in the container to about the background level.
2. **When using radioactive materials**, established rules and regulations must be followed. No source should be allowed to come into contact with the skin.

- Sources should be transferred using handling tools such as long-handled tongs or using robots with the material as far from the user as possible.
- Liquid and gas sources and solids in powder form should be in sealed containers. This is to ensure radioactive gas cannot be breathed in and radioactive liquid cannot be splashed on the skin or drunk.
- Radioactive sources should not be used for longer than is necessary. The longer a person is exposed to ionising radiation, the greater the dose of radiation received.

Summary questions

1. Explain why a source of α radiation is dangerous in the body but not as dangerous outside the body. *(3 marks)*

2. Explain why a radioactive source should be:
   a. kept in a lead-lined storage box when not in use *(2 marks)*
   b. out of its storage box for as little time as possible. *(2 marks)*

3. Describe two effects that ionising radiation has on living cells. *(2 marks)*

26.5 Radioactive decay
26.6 The theory of radioactive decay
Specification reference: 3.8.1.3

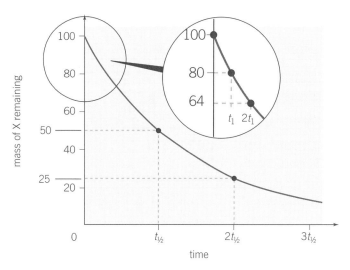

▲ **Figure 1** *A radioactive decay curve*

The half-life, $T_{1/2}$, of a radioactive isotope is the time taken for the mass of the isotope to decrease to half the initial mass. This is the same as the time taken for the number of nuclei or activity of the isotope to decrease to half the initial value.

The activity A of a radioactive isotope is the number of nuclei of the isotope that disintegrate per second. The unit of activity is the **becquerel (Bq),** where 1 Bq = 1 disintegration per second.

For a radioactive source of activity A that emits particles (or photons) of the same energy E, the energy transfer per second from the source = AE

Consider a sample of a radioactive isotope X that initially contains N_0 nuclei of the isotope. Let N represent the number of nuclei of X remaining at time t after the start. Suppose in time Δt, the number of nuclei that disintegrate is ΔN. The rate of disintegration $\dfrac{\Delta N}{\Delta t} = -\lambda N$ where λ is a constant referred to as the **decay constant**. The minus sign is because ΔN is a decrease.

The solution of the equation is $N = N_0 e^{-\lambda t}$, where e^x is the exponential function.

Note:
1. For a given radioactive isotope, its activity $A = (-)\dfrac{\Delta N}{\Delta t} = \lambda N$.
2. For a sample of a radioactive isotope, the mass m, the activity A, and the corrected count rate C of a detector at a fixed distance from the sample decreases from the initial values m_0 or A_0 or C_0 in accordance with the equation $X = X_0 e^{-\lambda t}$ where X represents m, A, or C (because these are all proportional to the number of nuclei, N, of the isotope).

> **Synoptic links**
>
> Reminder about molar mass and the Avogadro constant.
>
> For an element with a mass number A:
> - its molar mass M is its mass number in grams
> - mass m of the element contains $\dfrac{m}{M} N_A$ atoms where N_A is the Avogadro constant.
>
> See Topic 20.2, The ideal gas law, for more information.

> **Revision tip**
>
> In n half-lives of a certain isotope, the number of nuclei = the initial number ÷ 2^n = initial number × 0.5^n.

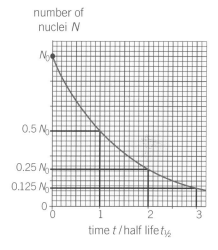

▲ **Figure 2** $N = N_0 e^{-\lambda t}$

> **Maths skill: Data analysis using natural logarithms**
>
> From the equation $N = N_0 e^{-\lambda t}$, taking natural logs of both sides gives $\ln N = \ln N_0 - \lambda t$. Therefore, a graph of $\ln N$ (or $\ln m$ or $\ln A$) against time t is a straight line with a negative gradient $-\lambda$ and a y-intercept equal to $\ln N_0$ (or $\ln m_0$ or $\ln A_0$).

Radioactivity

The decay constant

The decay constant λ is the probability of an individual nucleus decaying per second $= \left(\dfrac{\Delta N}{N \Delta t}\right)$.

The half-life $T_{1/2}$ is related to the decay constant λ according to the equation

$$T_{1/2} = \dfrac{\ln 2}{\lambda}$$

The unit of decay constant is s^{-1}. A large value of λ means fast decay and short half-life.

Summary questions

$N_A = 6.02 \times 10^{23}\,mol^{-1}$, $1\,MeV = 1.6 \times 10^{-13}\,J$

1 A freshly prepared sample of a radioactive isotope X contains 1.8×10^{15} atoms of the isotope. The half-life of the isotope is 2.0 hours. Calculate the number of atoms of this isotope remaining after:
   a 8.0 h *(2 marks)*
   b 3.0 h. *(2 marks)*

2 a Explain what is meant by the decay constant of a radioactive isotope. *(1 mark)*
   b The isotope $^{226}_{88}Ra$ has a half-life of 1620 years. For an initial mass of 1.0 kg of this isotope, calculate:
      i   the initial activity of this mass *(3 marks)*
      ii the mass of this isotope remaining after 3000 years *(4 marks)*
      iii how many atoms of the isotope will remain after 3000 years. *(1 mark)*

3 A fresh sample of a radioactive isotope has an initial activity of 40 kBq. After 24 h, its activity has decreased to 36 kBq. Calculate:
   a the decay constant of this isotope *(3 marks)*
   b its half-life *(2 marks)*
   c its activity after 120 h. *(3 marks)*

26.7 Radioactive isotopes in use

Specification reference: 3.8.1.2; 3.8.1.3

Radioactive dating

A radioactive isotope in an ancient object can be used to determine the age of the object by comparing:

- the activity per unit mass of the isotope with the activity per unit mass of the same isotope in a recently-formed object, or
- the ratio of the number of atoms of the radioactive isotope to the number of atoms of the isotope it decays into, provided the second isotope is stable.

Radioactive tracers

The radioactive isotope(s) in the tracer should:

- have a half-life which is long enough for the necessary measurements to be made and short enough to decay quickly after use
- emit β radiation or γ radiation so it can be detected outside the flow path.

Examples of industrial uses of radioactivity

Thickness monitoring

Figure 1 shows how metal foil is made using a continuous production line. The source used is a β⁻ emitter with a long half-life. α radiation would be absorbed completely by the foil and γ radiation would pass straight through without absorption.

if the radiation reaching the detector changes the detector makes the rollers move further apart or closer together

▲ **Figure 1** *The manufacture of metal foil*

Power sources for remote devices

Remote devices can be powered using a radioactive isotope. The isotope needs to have a reasonably long half-life but a very long half-life may require too much mass to generate the necessary power.

Summary questions

$N_A = 6.02 \times 10^{23}$ mol^{-1}, 1 MeV = 1.6×10^{-13} J

1. The uranium isotope $^{238}_{92}$U decays in several stages with an effective half-life of 4500 million years to form the lead isotope $^{206}_{82}$Pb. An ancient rock was found to contain 6.4 times as many $^{206}_{82}$Pb nuclei as $^{238}_{92}$U nuclei.
   a. Estimate the proportion of $^{238}_{92}$U in the rock now to when it formed. *(1 mark)*
   b. Use your estimate to determine the age of the rock. *(4 marks)*

2. A cardiac pacemaker is a device used to ensure that a faulty heart beats at a suitable rate. The electrical energy in a cardiac pacemaker is obtained from the plutonium isotope $^{238}_{94}$Pu which has a half-life of 88 years and emits α particles of energy 5.6 MeV. The initial mass of the isotope in the pacemaker is 0.20 grams.
   a. Calculate the initial activity of the radioactive source. *(3 marks)*
   b. Calculate the energy per second released from the source. *(2 marks)*

Revision tip

The choice of a radioactive isotope for a particular application is determined by:

- the half-life and the type of radiation needed
- whether or not a stable product is needed.

26.8 More about decay modes

Specification reference: 3.8.1.4

▲ Figure 1 The N–Z graph

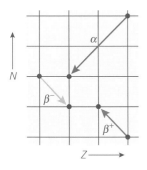

▲ Figure 2 N–Z changes

Synoptic link

β⁻ emitters can be manufactured by bombarding stable isotopes with neutrons. β⁺ emitters can only be produced by bombarding stable isotopes with protons. The protons need to have sufficient kinetic energy to overcome coulomb repulsion from the nucleus.

The N–Z graph

Figure 1 shows the N–Z graph of the neutron number N against the proton number Z for all known isotopes. Stable nuclei lie along a belt curving upwards with an increasing neutron–proton ratio from the origin to $N = 120$, $Z = 80$ approximately.

- For light isotopes (Z from 0 to no more than 20), the stable nuclei follow the straight line $N = Z$.

- As Z increases beyond about 20, stable nuclei have more neutrons than protons. The neutron/proton ratio increases. The extra neutrons help to bind the nucleons together without introducing repulsive electrostatic forces as more protons would do.

- α emitters occur beyond about $Z = 60$, most of them with more than 80 protons and 120 neutrons. The largest stable nuclide is $^{209}_{83}\text{Bi}$. Most α-emitting nuclides are above this.

- β⁻ emitters occur to the left of the stability belt where the isotopes are neutron-rich compared to stable isotopes.

- β⁺ emitters occur to the right of the stability belt where the isotopes are proton-rich compared to stable isotopes.

The change that takes place when an unstable nucleus becomes stable or less unstable can be represented on the N–Z graph as shown in Figure 2.

Radioactive series

An unstable nucleus, before it becomes stable, may undergo a series of changes in which each change involves an emission of an α particle or a β particle. Any radioactive series may be represented on the N–Z graph by a sequence of 'decay arrows'.

Nuclear energy levels

After an unstable nucleus emits an α or a β particle or undergoes electron capture, it might emit a γ photon in order to lose energy. This happens if the 'daughter' nucleus is formed in an **excited state** after it emits an α or a β particle or undergoes electron capture. The excited nucleus moves to its lowest energy state, its **ground state**, either directly or via one or more lower-energy excited states.

The technetium generator

Some radioactive isotopes such as the technetium isotope $^{99}_{43}\text{Tc}$ stay in the excited state long enough to be separated from the parent isotope. Such a long-lived excited state is said to be a **metastable state**. The technetium isotope $^{99}_{43}\text{Tc}$ in a metastable state (represented by the symbol $^{99}_{43}\text{Tc}^m$) forms after β⁻ emission from nuclei of the molybdenum isotope $^{99}_{42}\text{Mo}$ which has a half-life of 67 h. See Figure 3.

▲ Figure 3 The metastable state of technetium $^{99}_{43}\text{Tc}$

$^{99}_{43}Tc^m$ has a half-life of 6 h and decays to the ground state by γ emission. The ground state of $^{99}_{43}Tc$ is a β⁻ emitter with a half-life of 500 000 years and it forms a stable product. Therefore, a sample of $^{99}_{43}Tc^m$ with no molybdenum present effectively emits only γ photons.

$^{99}_{43}Tc^m$ is used in medical diagnosis, for example:

- **monitoring blood flow** through the brain after a small quantity of sodium pertechnate solution is administered intravenously
- **the γ camera** is designed to 'image' internal organs and bones by detecting γ radiation from sites in the body where a γ⁻ emitting isotope such as $^{99}_{43}Tc^m$ nuclei is located. For example, bone deposits can be located using a phosphate tracer labelled with $^{99}_{43}Tc^m$.

▲ **Figure 4** *The technetium generator. The ion exchange column contains ammonium molybdenate exposed to neutron radiation several days earlier to make a significant number of the molybdenum nuclei unstable. When a solution of sodium chloride is passed through the column, some of the chlorine ions exchange with pertechnate ions but not with molybdenate ions so the solution that emerges contains $^{99}_{43}Tc^m$ nuclei*

Summary questions

1. **a** Sketch an N–Z graph to show how N varies with Z for the known stable isotopes. *(2 marks)*
   b Show on your diagram possible locations of an isotope that is:
   i an α-emitter
   ii a β⁻ emitter
   iii a β⁺ emitter. *(3 marks)*

2. A nucleus of uranium $^{238}_{92}U$ decays to form a stable nucleus of lead $^{206}_{82}Pb$ by emitting 8 alpha particles and a number of β⁻ particles.
   a Determine the number of β⁻ particles it emits. *(2 marks)*
   b Use your N-Z chart to explain why no β⁺ particles are emitted. *(3 marks)*

3. The neon isotope $^{24}_{10}Ne$ releases 2.45 MeV of energy when it decays to form an isotope of sodium (Na). In the process, it releases β⁻ particles of maximum energy 1.98 MeV and 1.10 MeV and γ photons of energies 0.47 MeV and 0.88 MeV.
   a Construct an energy level diagram to show the possible ways in which a $^{24}_{10}Ne$ nucleus could decay. *(4 marks)*
   b Write the equation to represent the β⁻ decay of $^{24}_{10}Ne$. *(2 marks)*

26.9 Nuclear radius

Specification reference: 3.8.1.5

▲ **Figure 1** *High-energy electron diffraction*

Revision tip

1. For high-energy electrons their de Broglie wavelength $\lambda = \dfrac{hc}{E}$, where E is the energy of the electrons.
2. The angle θ_{min} depends on the radius R of the nucleus in accordance with the equation $R\sin\theta_{min} = 0.61\lambda$. You do *not* need to know the proof of this equation.

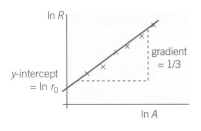

▲ **Figure 2** *Nuclear radius graphs*

Synoptic link

See Topic 1.2, Stable and unstable nuclei. The strong nuclear force has a range of about 2–3 fm which means the force only acts between 'nearest neighbour' neutrons and protons. So the density of a nucleus does not depend on its size.

High-energy electron diffraction

When a narrow beam of high-energy electrons is directed at a thin solid sample of an element, the incident electrons are diffracted by the nuclei of the atoms in the foil. Measurements of the number of electrons per second diffracted through different angles show that as the angle of diffraction θ is increased, the number of electrons per second diffracted into the detector decreases then increases slightly then decreases again. Scattering of the beam electrons by the nuclei due to their charge causes the intensity to decrease as θ increases. Diffraction of the beam electrons by each nucleus causes intensity maxima and minima to be superimposed on the effect above. The angle of the first minimum from the centre, θ_{min}, is measured and used to calculate the radius of the nucleus, provided the wavelength of the incident electrons is known.

Dependence of nuclear radius on nucleon number

Using samples of different elements, the radius R of different nuclides can be measured. It can be shown graphically that R depends on mass number A according to $R = r_0 A^{\frac{1}{3}}$, where the constant $r_0 = 1.05$ fm. A graph of $\ln R$ against $\ln A$ gives a straight line with a gradient of $\dfrac{1}{3}$ which confirms $R = r_0 A^{\frac{1}{3}}$ and gives a y-intercept equal to $\ln r_0$. A graph of R against $A^{\frac{1}{3}}$ gives a straight line through the origin with a gradient equal to r_0. Plotting this graph gives an accurate value of r_0.

Nuclear density

Assuming the nucleus is spherical:

- its volume $V = \dfrac{4}{3}\pi R^3 = \dfrac{4}{3}\pi (r_0 A^{\frac{1}{3}})^3 = \dfrac{4}{3}\pi r_0^3 A$
- its mass in atomic mass units = Au where 1u = 1 atomic mass unit = 1.661×10^{-27} kg.

Therefore, its density $\rho = \dfrac{A\text{u}}{\frac{4}{3}\pi r_0^3 A} = \dfrac{1\text{u}}{\frac{4}{3}\pi r_0^3}$ which is independent of A.

Summary questions

$r_0 = 1.05$ fm, 1u $= 1.661 \times 10^{-27}$ kg

1. A beam of high-energy electrons directed at a target is diffracted by the nuclei in the target.

   a. The angle of diffraction of the first minimum was 39° when the energy of the electrons was 310 MeV. Calculate the de Broglie wavelength of the electrons and use $R \sin \theta_{min} = 0.61\lambda$ to calculate the radius of the target nuclei. *(3 marks)*

2. In the high-energy electron diffraction experiment in **Q1**, the angle of diffraction was measured to within $\pm 2.0°$. Estimate the uncertainty in the radius of the target nucleus due to this measurement. *(2 marks)*

3. Use the equation $R = r_0 A^{\frac{1}{3}}$ to calculate the radius of a $^{28}_{14}$Si nucleus and hence determine its density. *(3 marks)*

Chapter 26 Practice questions

1. A polonium nucleus $^{218}_{84}$Po undergoes a series of α and β decays before it forms a stable lead nucleus $^{206}_{82}$Pb. Which one of the following alternatives, **A–D**, gives the total number of decays?

   A 4   **B** 5   **C** 6   **D** 7   *(1 mark)*

2. The half-life of a radioactive isotope X is 10 hours. A pure sample of X is prepared. How long will it take for the sample to contain three times as many decayed atoms as undecayed atoms of X?

   A 10 hours   **B** 20 hours   **C** 30 hours   **D** 40 hours   *(1 mark)*

3. Two radioactive isotopes P and Q have half-lives of 50 s and 100 s respectively. Separate samples of the two isotopes are prepared at the same time with the same initial activity. The isotopes decay into stable isotopes R and S. Which one of the following alternatives, **A–D**, gives the ratio $\frac{\text{the number of atoms of Q}}{\text{the number of atoms of P}}$ after 200 s?

   A 4   **B** 8   **C** 12   **D** 16   *(1 mark)*

4. The following count rate measurements were made of the radiation from a radioactive isotope at a fixed distance from a Geiger tube connected to a counter. The average background count was 25 counts per minute.

Time t / s	0	30	60	90	120	150	180
Count rate /s	11.7	8.4	6.6	5.2	4.2	3.0	2.5
Corrected count rate C /s							

   a Calculate the corrected count rate C and copy and complete the last row of the table. *(1 mark)*

   b Plot a graph of ln C on the y-axis against time t on the x-axis. *(3 marks)*

   c i Use your graph to determine the half-life of the radioactive isotope. *(3 marks)*

   ii Explain why the Geiger tube needed to be at the same distance from the isotope throughout the measurements. *(2 marks)*

5. a A beam of monoenergetic α particles is directed normally at a thin metal foil in an α-scattering experiment.

   i Explain why a small proportion of the α particles are deflected through angles of more than 90°. *(3 marks)*

   ii Explain why multiple scattering of an individual α particle is unlikely. *(3 marks)*

   b A narrow beam of 4.5 MeV α particles is directed normally at a thin copper foil. Estimate the least distance of approach of an α particle to a copper $^{63}_{29}$Cu nucleus. *(3 marks)*

6. A small source of γ radiation inside a sealed container was placed at different distances d from a Geiger tube and the count rate was measured at each distance. The results corrected for background radiation are shown in the table below.

Distance d / mm	102	136	170	202	245	305	385
Corrected count rate C / s^{-1}	218	126	73	57	40	27	15

   Assuming the source in the container is at a distance d_0 from the front of the container, the distance from the source to the Geiger tube is $d + d_0$, so the corrected count rate C is inversely proportional to $(d + d_0)^2$ in accordance with the inverse-square law.

Radioactivity

    a Show that $d = \dfrac{a}{\sqrt{C}} + b$, where a and b are constants. *(2 marks)*

    b **i** Tabulate the values of $\dfrac{1}{\sqrt{C}}$ and d. *(2 marks)*

        ii Plot a suitable graph to find out if the inverse-square law applies to these results. *(3 marks)*

    c Use your graph to determine the distance d_0. *(1 mark)*

7 A sample of the stable copper isotope $^{63}_{29}\text{Cu}$ is placed in the core of a nuclear reactor where the copper atoms are bombarded by neutrons. As a result, the sample becomes radioactive due to some of its nuclei each absorbing a neutron.

    a Write down an equation to represent a copper nucleus absorbing a neutron. *(1 mark)*

    b **i** What type of radiation is emitted by the copper nucleus after it has absorbed a neutron? Give a reason for your answer. *(3 marks)*

        ii After absorbing a neutron, the copper nucleus emits radiation and forms a stable zinc (Zn) nucleus. Write down an equation to represent the change that takes place as a result of the emission of the radiation in part **b i**. *(2 marks)*

    c Discuss whether or not the sample would become radioactive if it were to be bombarded by protons instead of neutrons. *(2 marks)*

8 When a radioactive source near a Geiger tube was moved gradually away from the tube, the count rate due to the source suddenly dropped significantly.

    a Explain why the count rate suddenly dropped. *(2 marks)*

    b The count rate did not drop to zero. Describe a further test you would do to determine the type(s) of radiation from the source still being detected. *(4 marks)*

27.1 Energy and mass

Specification reference: 3.8.1.6

Energy and the strong nuclear force

- The strong nuclear force acts between any two protons or neutrons when they are closer than about 3 to 4 femtometres. The force is attractive at distances greater than 0.5 fm (10^{-15} m) and repulsive at shorter distances.
- The strength of the strong nuclear force is sufficient to overcome the Coulomb repulsion between 2 protons at a separation of about 1 fm which is about 200 N. So the strong nuclear force is at least 200 N.
- Nuclear changes involve energies of the order of MeV ($\approx 200\,\text{N} \times 3 \times 10^{-15}\,\text{m} = 6 \times 10^{-13}\,\text{J} \approx 3\text{–}4\,\text{MeV}$ as $1\,\text{MeV} = 1.6 \times 10^{-13}\,\text{J}$).

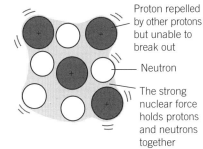

▲ **Figure 1** *The strong nuclear force*

Energy changes in reactions

Reactions on a nuclear or sub-nuclear scale involve measurable changes of mass. For a difference Δm in the total mass of all the particles involved before and after a reaction, the energy transferred

$$Q = \Delta m c^2$$

1. In **α decay**, the energy released is shared as kinetic energy between the α particle and the nucleus. The two particles move in opposite directions with equal and opposite momentum. So the energy released is shared between the α particle and the nucleus in inverse proportion to their masses.

2. In **β decay**, the energy released is shared in variable proportions between the β particle, the nucleus, and the neutrino or antineutrino released in the decay. The maximum kinetic energy of the β particle is very slightly less than the energy released in the decay because of recoil of the nucleus.

3. In **electron capture**, the nucleus emits a neutrino which causes the nucleus to recoil. The energy released is shared between the nucleus and the neutrino which carries away most of the energy released. The atom also emits an X-ray photon when the inner-shell vacancy due to the electron capture is filled.

Revision tips

1. When calculating Q in beta decay, assume the mass of the neutrino is negligible.
2. To calculate the energy released in a reaction in MeV, calculate the mass difference in u then multiply by 931.5 MeV/u.

Revision tip

If the mass of each atom is given instead of the mass of its nucleus, calculate the mass of each nucleus by subtracting the mass of the electrons ($= Zm_e$) in the atom from the mass of each atom.

Revision tip

Einstein showed that the mass m of an object increases (or decreases) when it gains (or loses) energy E in accordance with the equation $E = mc^2$. This equation applies to all energy changes.

Worked example

The radon isotope $^{220}_{86}\text{Rn}$ emits α particles and decays to form the polonium isotope $^{220}_{86}\text{Rn}$.

Write down an equation to represent this process and calculate the energy released when a $^{220}_{86}\text{Rn}$ nucleus emits an α particle.

mass of $^{220}_{86}\text{Rn}$ nucleus = 219.964 18 u

mass of $^{216}_{84}\text{Po}$ nucleus = 215.955 76 u

mass of α particle = 4.001 50 u

1 u is equivalent to 931.5 MeV

$^{220}_{86}\text{Rn} \rightarrow \,^{4}_{2}\alpha + \,^{216}_{84}\text{Po}$ (+ energy released Q)

mass difference = total initial mass − total final mass = 219.964 18 − (215.955 76 + 4.001 50) = 6.92×10^{-3} u

energy released Q = mass difference in u × 931.5 MeV/u = 6.45 MeV

Nuclear energy

Summary questions

1 u = 931.5 MeV $g = 9.81 \text{ m s}^{-2}$

1 A carbon $^{14}_{6}\text{C}$ nucleus decays to form a stable nucleus of nitrogen (N) and emits a β⁻ particle and an antineutrino in the process.
   a Write down an equation to represent this process. *(2 marks)*
   b Calculate the energy released.
   mass of $^{14}_{6}\text{C}$ nucleus = 13.999 95 u,
   mass of the nitrogen
   nucleus = 13.999 23 u
   mass of β⁻ particle = 0.000 55 u *(3 marks)*

2 The argon isotope $^{37}_{18}\text{Ar}$ decays through electron capture to form the stable isotope of chlorine (Cl).
   a Write down an equation to represent this process. *(4 marks)*
   b Calculate the energy released.
   mass of $^{37}_{18}\text{Ar}$ nucleus = 36.956 89 u,
   mass of chlorine nucleus
   nucleus formed = 36.956 57 u
   mass of electron = 0.000 55 u *(3 marks)*

3 The americium isotope $^{241}_{95}\text{Am}$ emits α particles and decays to form an isotope of neptunium (Np).
   a Write down an equation to represent this process. *(2 marks)*
   b Calculate the energy released. *(4 marks)*
   atomic mass of $^{241}_{95}\text{Am}$ = 241.056 69 u,
   atomic mass of neptunium
   isotope = 237.048 03 u
   mass of α particle = 4.001 50 u, mass of electron = 0.000 55 u

27.2 Binding energy
Specification reference: 3.8.1.6

Binding energy

When a nucleus forms from separate neutrons and protons, energy is released as the strong nuclear force does work pulling the nucleons together. Because energy is released when a nucleus forms from separate neutrons and protons, the mass of a nucleus is less than the mass of the separated nucleons.

The mass defect Δm of a nucleus is defined as the difference between the mass of the separated nucleons and the mass of the nucleus.

For a nucleus $^A_Z X$ of mass M_{NUC}, **its mass defect $\Delta m = Zm_p + (A - Z)m_n - M_{NUC}$**

where m_p and m_N represent the masses of the proton and the neutron respectively.

The binding energy of the nucleus is the work that must be done to separate a nucleus into its constituent neutrons and protons.

For a nucleus which has a mass defect Δm, **its binding energy = Δmc^2**

> **Revision tip**
> The energy corresponding to a mass of 1 u is $1.661 \times 10^{-27} \times (3.00 \times 10^8)^2$ J = 931.5 MeV. If you are given the mass of the nucleus in kilograms, you can convert this to atomic mass units and use the method above to calculate the mass defect. The values of the mass of the proton and the neutron are given in atomic mass units in the data sheet.
>
> **Note:** The atomic mass unit, 1 u = 1.661×10^{-27} kg, is defined as $\frac{1}{12}$th of the mass of an atom of the carbon isotope $^{12}_6 C$.

The binding energy curve

The **binding energy per nucleon** of a nucleus is the average work done per nucleon to remove all the **nucleons** (protons and neutrons) from a nucleus – it is therefore a measure of the stability of a nucleus. If the binding energies per nucleon of two different nuclides are compared, the nucleus with more binding energy per nucleon is the more stable of the two nuclei.

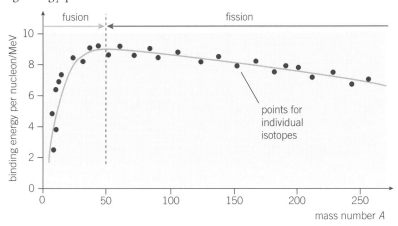

▲ **Figure 1** *Binding energy per nucleon for all known nuclides*

Figure 1 shows a graph of the binding energy per nucleon against mass number A for all the known nuclides. This graph is a curve which has a maximum value of 8.7 MeV per nucleon between $A = 50$ and $A = 60$. Nuclei with mass numbers in this range are the most stable nuclei.

When energy is released by a nucleus, its binding energy per nucleon increases. Energy is released in:

- **nuclear fission**, the process in which a large unstable nucleus splits into two fragments which are more stable than the original nucleus. The binding energy per nucleon increases in this process, as shown in Figure 1.

- **nuclear fusion**, the process of making small nuclei fuse together to form a larger nucleus. The product nucleus has more binding energy per nucleon than the smaller nuclei. So the binding energy per nucleon also increases in this process, provided the nucleon number of the product nucleus is no greater than about 50.

Nuclear energy

Worked example

The mass of a nucleus M_{NUC} of the iron isotope $^{56}_{26}$Fe is 55.920 66 u. Calculate the binding energy of this nucleus in MeV.

mass of a proton, $m_p = 1.007\,28$ u

mass of a neutron, $m_n = 1.008\,67$ u

1 u is equivalent to 931.5 MeV

Mass defect $\Delta m = 26 m_p + (56 - 26) m_n - M_{NUC} = 0.528\,72$ u

Therefore, binding energy $= 0.528\,72$ u $\times$ 931.5 MeV/u $= 493$ MeV

Binding energy per nucleon $= \dfrac{493\,\text{MeV}}{56} = 8.8$ MeV

Summary questions

mass of a proton $m_p = 1.007\,28$ u, mass of a neutron $m_n = 1.008\,67$ u, 1 u is equivalent to 931.5 MeV

1. **a** Explain what is meant by the binding energy of a nucleus. *(2 marks)*
   b Describe and explain how the binding energy per nucleon of a heavy nucleus changes when it emits an alpha particle. *(2 marks)*

2. Calculate the binding energy per nucleon, in MeV per nucleon, of:
   a a $^3_2$He nucleus *(3 marks)*
   b a $^{238}_{92}$U nucleus. *(3 marks)*

   mass of an $^3_2$He nucleus $= 3.015\,975$ u
   mass of a $^{238}_{92}$U nucleus $= 238.050\,74$ u

3. **a** Write down an equation to show the fusion reaction that occurs when a proton and a $^2_1$H nucleus fuse together to form a $^3_2$He nucleus. *(2 marks)*
   b Calculate the energy released in the above reaction. *(3 marks)*

   mass of $^2_1$H nucleus $= 2.013\,55$ u   mass of $^3_2$He nucleus $= 3.014\,93$ u

27.3 Fission and fusion

Specification reference: 3.8.1.7; 3.8.1.8

Induced fission

Fission of a nucleus occurs when a nucleus splits into two approximately equal fragments. This happens when the uranium isotope $^{235}_{92}U$ is bombarded with neutrons. This process is known as **induced fission**.

Each fission event releases energy and two or three neutrons referred to as **fission neutrons**. The neutrons are each capable of causing a further fission event as a result of a collision with another $^{235}_{92}U$ nucleus. A **chain reaction** is therefore possible in which fission neutrons produce further fission events which release fission neutrons and cause further fission events and so on. Complete fission of $^{235}_{92}U$ would release about a million times more than the energy released as a result of burning a similar mass of fossil fuel.

Many fission products are possible when a fission event occurs. The following equation shows a possible fission event in which a $^{235}_{92}U$ nucleus is split into a barium $^{144}_{56}Ba$ nucleus and a krypton $^{90}_{36}Kr$ nucleus and two neutrons are released.

$$^{235}_{92}U + ^{1}_{0}n \rightarrow ^{144}_{56}Ba + ^{90}_{36}Kr + 2^{1}_{0}n + \text{energy released, } Q$$

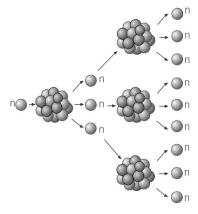

▲ **Figure 1** *A chain reaction in a nuclear reactor*

Nuclear fusion

Fusion takes place when two nuclei combine to form a bigger nucleus. The binding energy per nucleon of the product nucleus is greater than of the initial nuclei. As a result, energy is released equal to the increase of binding energy.

Nuclear fusion can only take place if the two nuclei that are to be combined collide at high speed. This is necessary to overcome the electrostatic repulsion between the two nuclei so they can become close enough to interact through the strong nuclear force. For example, the reaction below represents the fusion of a nucleus of deuterium (the hydrogen isotope $^{2}_{1}H$) and a nucleus of tritium (the hydrogen isotope $^{3}_{1}H$).

$$^{2}_{1}H + ^{3}_{1}H \rightarrow ^{4}_{2}He + ^{1}_{0}n + 17.6\,\text{MeV}$$

▲ **Figure 2** *Fusion of two protons*

> **Revision tip**
> 1 The energy released in a fission or fusion reaction may be calculated using $E = mc^2$ in the form $Q = \Delta mc^2$, where Δm is the difference between the total mass before and after the event.
> 2 In each type of reaction, the total number of neutrons and protons is unchanged but in a fusion reaction in which a β particle is emitted, a neutron changes into a proton or a proton changes into a neutron.
> $^{1}_{1}p + ^{1}_{1}p \rightarrow ^{2}_{1}H + ^{0}_{+1}\beta + 17.6\,\text{MeV}$

Fusion power

The temperature of the material containing the nuclei to be fused must be of the order of 10^8 K so that the nuclei are moving fast enough to fuse when they collide with each other. The material at this temperature is called a plasma as it consists of positive ions and unattached electrons.

A very strong magnetic field is used to confine the plasma gas so it doesn't make contact with the inside of its container and lose its energy. The plasma is heated by passing a very large current through it. At present, confinement can only be achieved for a relatively short duration during which time more power is released than is supplied to the plasma.

Nuclear energy

Summary questions

$1\,u \equiv 931\,\text{MeV}$

1. **a** Explain why the protons in a nucleus do not leave the nucleus even though a repulsive force acts between them due to their positive charge. *(2 marks)*
   b Explain why the mass of a nucleus is less than the mass of the separated protons and neutrons from which the nucleus is composed. *(3 marks)*

2. The equation below represents a reaction that takes place when a neutron collides with a nucleus of the uranium isotope $^{235}_{92}\text{U}$.

   $$^{235}_{92}\text{U} + ^{1}_{0}\text{n} \rightarrow\ ^{140}_{54}\text{Xe} + ^{93}_{38}\text{Sr} + 3\,^{1}_{0}\text{n} + \text{energy released},\ Q$$

   Calculate the energy, in MeV, released in this fission reaction. *(4 marks)*

   mass of $^{235}_{92}\text{U}$ nucleus 234.943 42 u

   mass of $^{140}_{54}\text{Xe}$ nucleus 139.891 92 u

   mass of $^{93}_{38}\text{Sr}$ nucleus 92.893 13 u   mass of a neutron 1.008 67 u

3. In a fusion reactor, a hydrogen $^{3}_{1}\text{H}$ nucleus is fused with a hydrogen $^{2}_{1}\text{H}$ nucleus to form a helium $^{4}_{2}\text{He}$ nucleus. In the process another particle X is released.
   a Write down an equation to represent this fusion reaction and to identify X. *(2 marks)*
   b Calculate the energy released in this reaction using the appropriate data below. You will need to use some or all of this data. *(4 marks)*

   mass of a β particle 0.000 55 u   mass of a proton 1.007 28 u
   mass of a neutron 1.008 67 u
   mass of $^{3}_{1}\text{H}$ nucleus 3.015 50 u   mass of $^{2}_{1}\text{H}$ nucleus 2.013 55 u
   mass of $^{4}_{2}\text{He}$ nucleus 4.001 50 u

27.4 The thermal nuclear reactor

Specification reference: 3.8.1.7; 3.8.1.8

Inside a nuclear reactor

A **thermal nuclear reactor** in a nuclear power station consists of a steel vessel known as the **reactor core** connected by steel pipes to a **heat exchanger**. The reactor core contains:

1. fuel rods spaced evenly in a **moderator** such as water under pressure or graphite
2. the moderator, which is necessary to slow the fission neutrons so they can cause further fission
3. **control rods** made of material such as boron which absorbs neutrons. The depth of the rods in the moderator is automatically adjusted so that the rate of fission events is constant
4. a **coolant** (e.g., water or carbon dioxide gas) that is pumped through steel pipes between the reactor core and the moderator so that it transfers energy from the reactor vessel to the heat exchanger. The coolant must flow easily and it must not be corrosive.

> **Revision tip**
> The fission neutrons transfer energy to the moderator atoms as a result of repeated elastic collisions with the atoms. In each collision, the total kinetic energy of the neutron and the atom is conserved as well as their total momentum. See Topic 27.4, The thermal nuclear reactor, for more about a simple mechanical model of the moderation process.

Fission facts

- The fuel rods contain enriched uranium mostly of the non-fissionable uranium isotope $^{238}_{92}U$, with about 2–3% of the uranium isotope $^{235}_{92}U$ which is fissionable. In comparison, natural uranium contains 99% U-238.
- The depth of the control rods in the core is automatically adjusted to keep the number of neutrons in the core constant so that exactly one fission neutron per fission event on average goes on to produce further fission. This condition keeps the rate of release of fission energy constant.
- The fission neutrons are slowed down by repeated collisions with the moderator atoms to increase the probability of induced fission. The process is most effective if the mass of the moderator atom is as close as possible to the mass of the neutron.
- For a chain reaction to occur, the mass of the fissile material (e.g., U-235) must be greater than a minimum mass, referred to as the **critical mass**. If the mass of fissile material is less than the critical mass needed, too many of the fission neutrons escape because the surface area to mass ratio of the material is too high.

> **Revision tip**
> Critical mass refers to the mass of the fissile material not the atomic mass.

▼ **Table 1** *Comparison of thermal reactors*

	Advanced gas-cooled reactor	Pressurised water reactor
Fuel	uranium oxide in stainless steel cans	uranium oxide in zirconium alloy cans
Moderator	graphite	water
Coolant	CO_2 gas	water
Coolant temperature/K	900	600
Typical power output/MW	1300	700

Safety features

1. The reactor core is a thick steel vessel that absorbs β radiation and some of the γ radiation and neutrons from the core.
2. The core is surrounded by very thick concrete walls which absorb the neutrons and γ radiation that escape from the reactor vessel.

Nuclear energy

3 Every reactor has an emergency shut-down system designed to insert the control rods fully into the core to stop fission completely.

4 The sealed fuel rods are inserted and removed from the reactor by means of remote handling devices. The rods are much more radioactive after removal than before because the used fuel rods emit β and γ radiation due to the many neutron-rich fission products that form.

Radioactive waste

High-level radioactive waste such as spent fuel rods contains many different radioactive isotopes, including fission fragments as well as unused uranium-235, uranium-238, and plutonium-239.

The spent fuel rods are removed by remote control and stored underwater in cooling ponds for up to a year because they continue to release heat due to radioactive decay.

The rods are then transferred in large steel casks to a reprocessing plant where the unused uranium and plutonium is then removed and stored in sealed containers for further possible use.

The rest of the material is radioactive waste and is stored for many years (as it contains long-lived radioactive isotopes) in sealed containers in deep trenches or in geologically-stable underground caverns. In some countries, high-level radioactive waste is vitrified by mixing it with molten glass and then stored as glass blocks in underground caverns.

Intermediate-level waste such as radioactive materials with low activity and containers of radioactive materials are sealed in drums that are encased in concrete and stored in specially constructed buildings with walls of reinforced concrete.

Low-level waste such as laboratory equipment and protective clothing is sealed in metal drums and buried in large trenches.

> ### Summary questions
>
> 1 a What is the function of the moderator in a nuclear reactor?
>   *(2 marks)*
>   b Describe how the moderator works. *(5 marks)*
>
> 2 Describe and explain how the control rods of a nuclear reactor are used when a nuclear reactor is:
>   a releasing energy at a constant rate *(3 marks)*
>   b shut down. *(4 marks)*
>
> 3 Explain why the spent fuel rods from a nuclear reactor are more radioactive after removal from the reactor than they were before they were used in the reactor. *(5 marks)*

Chapter 27 Practice questions

1 u ≡ 931.5 MeV

1. A nucleus $^A_Z X$ has a mass M. The mass of a proton is m_p and the mass of a neutron is m_n. The mass defect of the nucleus can be written as $rA + sZ + tM$. Which one of the following alternatives, **A–D**, gives the correct terms for r, s, and t? *(1 mark)*

	r	s	t
A	m_p	m_p+m_n	1
B	m_p	m_p+m_n	−1
C	m_p	m_p-m_n	1
D	m_p	m_p-m_n	−1

2. A nuclear power station has an electrical power of 1200 MW and an efficiency of 40%. It uses uranium fuel that contains the fissile isotope $^{235}_{92}$U. Each fission event releases approximately 200 MeV. Which one of the following alternatives, **A–D**, gives the nearest estimate of the number of fission events per second?

   A 10^{12} **B** 10^{14} **C** 10^{16} **D** 10^{20} *(1 mark)*

3. A $^{235}_{92}$U nucleus undergoes induced fission to form two nuclei X and Y with the emission of 2 fission neutrons. The mass number of X is P and its atomic number is Q. Which one of the following alternatives, **A–D**, gives the mass number and atomic number of Y? *(1 mark)*

	mass number	atomic number
A	233 − P	91 − Q
B	234 − P	91 − Q
C	233 − P	92 − Q
D	234 − P	92 − Q

4. The nuclide $^{64}_{29}$Cu can decay by beta plus (β⁺) emission to form a nickel (Ni) nucleus.

   a Copy and complete the equation below to represent this decay.

   $^{64}_{29}\text{Cu} \rightarrow \ ^{\ 0}_{\_}\beta + \ ^{\_}_{\_}\text{Ni} + \_ + Q$ *(2 marks)*

   b The energy, Q, released in the above decay is 0.66 MeV. Use this information and the data below to calculate the mass of the nickel nucleus in atomic mass units. *(3 marks)*

   mass of $^{64}_{29}$Cu nucleus = 63.913 84 u   mass of an electron = 0.000 55 u

   c Figure 1 represents the above change in terms of quarks. Identify the particles a, b, c, and d and copy and complete the table below.

	Name	Symbol
a		
b		
c		
d		

   (2 marks)

   ▲ Figure 1

   d State one similarity and one difference between a β⁺ particle and the particle in Figure 1 labelled:

   i   d

   ii c. *(4 marks)*

Nuclear energy

5 a i Sketch a graph to show how the binding energy per nucleon depends on the nucleon number *A* for all known nuclides. Show appropriate values on the axes of your graph. *(4 marks)*

ii Use your graph to explain why energy is released when two small nuclei are fused together in a fusion reaction. *(4 marks)*

b i The mass of a helium $^4_2$He nucleus is 4.00150 u. Use this information and the data below to calculate the binding energy per nucleon of a helium nucleus.

mass of a proton = 1.00728 u; mass of a neutron = 1.00867 u

(3 marks)

ii Mark and label the position of a $^4_2$He nucleus on your graph. *(1 mark)*

iii The binding energy of a $^2_1$H nucleus is about a sixth of the binding energy of an alpha particle. Explain why the emission of an alpha particle from a large unstable nucleus is much more likely than the emission of a $^2_1$H nucleus. *(3 marks)*

6 a Explain what is meant by induced fission. *(2 marks)*

b i Describe what is meant by a chain reaction of fission events. *(5 marks)*

ii Explain why there needs to be a minimum mass of fissile material for a chain reaction to occur. *(4 marks)*

c In addition to fuel rods, a thermal nuclear reactor contains control rods and a coolant.

i State the function of the control rods, describe how they are used, and give an example of a suitable material from which they are manufactured. *(5 marks)*

ii The moderator is used to reduce the kinetic energy of the neutrons produced by the fission reactions. State the reason why this is necessary and explain why the process of reducing the kinetic energy of a fission neutron is more effective if the atomic mass of the moderator atoms is as close as possible to the mass of a neutron. *(5 marks)*

iii State the essential properties of the coolant and describe how it is used to transfer energy from the nuclear reactor. *(6 marks)*

7 a The polonium nucleus $^{210}_{84}$Po emits an alpha particle and forms a lead (Pb) nucleus.

i Copy and complete the equation below to represent this reaction.

$$^{210}_{84}\text{Po} \rightarrow \,\underline{}\text{Pb} + \underline{}\alpha + Q$$

(2 marks)

ii Use the data below to calculate the energy *Q* released in MeV in this reaction.

mass of $^{210}_{84}$Po nucleus = 209.93675 u

mass of lead nucleus formed = 205.92945 u

mass of alpha particle = 4.00150 u *(3 marks)*

b i The energy released is shared as kinetic energy between the lead nucleus and the alpha particle. Explain why the alpha particle does not take all the energy released in the reaction. *(2 marks)*

ii Calculate the kinetic energy of the alpha particle immediately after it was emitted by the nucleus. *(6 marks)*

Astrophysics

The notes below highlight key points in your option. Use these revision notes alongside the notes for your option in your AQA A Level Physics Student Book and the detailed notes on the option in the corresponding Kerboodle resources. Use the questions provided here and in the Kerboodle resources to check your knowledge and skills.

CHAPTER 1 TELESCOPES

1.1 Lenses

The **lens formula** $\frac{1}{u} + \frac{1}{v} = \frac{1}{f}$ where u is the distance from the lens to the object, v is the distance from the lens to the image, and f is the focal length of the lens.

Linear magnification $= \frac{\text{height of object}}{\text{height of image}} = \frac{v}{u}$

Summary questions

1. An object of height 18.0 mm is placed perpendicular to the axis of a converging lens of focal length 0.120 m at a distance of 0.080 m from the lens.
   a Calculate the image distance and the height of the image formed by the lens. *(2 marks)*
   b Describe the image. *(1 mark)*

Revision tip
Make sure you can draw a ray diagram to scale to locate an image formed by a converging lens.

1.2–1.5 Telescope design

The refracting telescope

In normal adjustment:

- the image of a distant object is at infinity
- the distance between the two lenses $= f_o + f_e$
- the angular magnification $M = \frac{f_o}{f_e}$.

The Cassegrain telescope has a parabolic concave mirror as the objective and a small 'secondary' convex mirror to reflect light into the eyepiece.

Revision tip
Make sure you can draw a ray diagram for a refracting telescope in normal adjustment and for a Cassegrain telescope.

Summary questions

1. a A telescope consists of two converging lenses of focal lengths 60 mm and 480 mm. It is used to view a distant object that subtends an angle of 0.150° to the telescope. Calculate:
   i the angular magnification of the telescope *(1 mark)*
   ii the angle subtended by the virtual image seen by the viewer. *(1 mark)*
   b Explain why a Cassegrain reflector with a parabolic mirror does not cause spherical or chromatic aberration whereas a refractor telescope does. *(2 marks)*
   c Two nearby stars are observed using a telescope with an objective of diameter 80 mm. Assuming light from them has an average wavelength of 500 nm, calculate their angular separation in degrees. *(2 marks)*

Revision tip
1. To resolve two stars that are almost in the same direction, their angular separation $\geq \frac{\lambda}{D}$ where D is the telescope diameter.
2. Advantages of a CCD:
   - Quantum efficiency of a CCD pixel ≈70% compared with ≈ 4% for photographic film.
   - Wavelength sensitivity (≈100 nm to 1100 nm) > the human eye (≈ 350 nm to 650 nm).
   - Images recorded continuously, stored, and processed electronically.

81

Astrophysics

> **Revision tip**
> Make sure you know the difference between apparent magnitude m and absolute magnitude M and how to use the equation
> $m - M = 5 \log \dfrac{d}{10}$

> **Revision tip**
> Make sure you know how to apply **Wien's Law** and **Stefan's Law** to determine the temperature and the power output of a star. Remember temperature is always in kelvins.

CHAPTER 2 SURVEYING THE STARS
2.1 Star magnitudes
2.2 Classifying stars
Star distances and magnitudes

The **parallax angle** θ of a star is the angle subtended by the star to the line between the Sun and the Earth when the line is perpendicular to the line of sight to the star.

The distance, in parsecs, from a star to the Sun $= \dfrac{1}{\theta(\text{in arcseconds})}$

Note: 1 arc second = $\dfrac{1}{3600}$ degree. Also 1 parsec = 3.26 light years.

Stellar spectral classes

The spectrum of light from a star is used to classify it. Make sure you know the main star classes (OBAFGKM), their approximate temperatures and colour. Hydrogen absorption lines (only in O, B, and A stars) are due to excitations from the $n = 2$ energy level (The Balmer lines).

> **Summary questions**
>
> 1 a State what is meant by the absolute magnitude of a star. *(1 mark)*
>   b A star has an apparent magnitude of + 4.8 and its distance from Earth is 34 pc. Calculate its absolute magnitude. *(2 marks)*

2.3 The Hertzsprung–Russell (HR) diagram

Figure 1 shows the HR diagram. Make sure you can draw and the label the HR diagram and that you know the main characteristics of each type of star shown on the diagram.

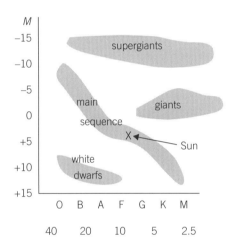

▲ **Figure 1** *The Hertzsprung–Russell diagram*

> **Key terms**
>
> **Formation of a star:** Dust and gas clouds → protostar due to gravity; core heats up due to nuclear fusion of hydrogen into helium; outer layers of the protostar (the photosphere) heat up and emit light.
>
> **Main sequence:** Inward gravitational attraction balanced by the outward radiation pressure. The star has a constant luminosity; hydrogen in the core → helium.
>
> **Red giants:** Core collapses as hydrogen all converted to helium; star becomes a giant or a supergiant as the outer layers of the star expand and cool. A star of:
> - mass less than 8 solar masses → a white dwarf
> - higher mass → **a supergiant** which explodes as a supernova.
>
> **White dwarfs:** Cools and its core contracts, throwing off its outer layers. If its mass is:
> - less than 1.4 solar masses, it becomes stable, as **a white dwarf** which cools and eventually fades out
> - greater than 1.4 solar masses, it explodes as a **supernova**.

> **Summary questions**
>
> Two stars P and Q have surface temperatures of 3000 K and 10 000 K respectively and absolute magnitude −4, and +4 respectively.
> 1 a State which star has the greater power output and state the further evolutionary stages of each star. *(4 marks)*
>   b Calculate the ratio of the diameter of P relative to Q. *(4 marks)*

2.4 Supernovas, neutron stars, and black holes

In a supernova event, the core of a star collapses and becomes denser and denser until it is rigid. The matter collapsing onto the core rebounds off the core, sending matter outwards into space.

Type 1a supernovas each have a peak luminosity about $M = -18$ and a strong silicon absorption line. They are used to find the distance to their host galaxy.

Neutron stars and black holes

A neutron star is the core of a supernova after all the surrounding matter has been thrown off into space.

A black hole is an object so dense that not even light can escape from it. The main physical property of a black hole is its mass.

The event horizon of a black hole is a sphere surrounding the black hole from which nothing can ever emerge. The radius of its event horizon (its Schwarzschild radius) $R_s = \frac{2GM}{c^2}$.

Supermassive black holes of enormous mass are thought to exist at the centres of many galaxies where they pull in gas and stars from their 'host' galaxy.

CHAPTER 3 COSMOLOGY
3.1 The Doppler effect

For a source of light moving at speed v relative to an observer,

$$\text{the change of wavelength of the light } \Delta\lambda = \pm\frac{v}{c}\lambda$$

The formula above holds provided $v \ll$ the speed of light c.

The **Doppler shift**, z, in frequency (or wavelength) is the fractional change $\frac{\Delta f}{f}$ (or $\frac{\Delta\lambda}{\lambda}$).

For a star or galaxy that is moving:

- away from the Earth, a **red shift** is observed (i.e., an increase of wavelength)
- towards the Earth, a **blue shift** is observed (i.e., a decrease of wavelength).

3.2 Hubble's Law and beyond
3.3 Quasars

Galaxies

Hubble's Law states that the distant galaxies are receding from us and their speed of recession v is directly proportional to the distance, d, to it.

$$v = Hd$$

where H, the constant of proportionality, is referred to as the Hubble constant.

Notes:

1. The red shift of a distant galaxy is used to calculate its speed of recession.
2. Its distance from Earth is determined by observing the period of individual Cepheid variable stars in the galaxy and using the known relationship between the period of a Cepheid variable and its mean absolute magnitude.

The Big Bang theory

The Big Bang theory states that the Universe was created in a massive 'primordial' explosion. The main evidence for the Big Bang theory is:

- **cosmic microwave background radiation** (CMBR), which is due to high-energy radiation created in the Big Bang that has been travelling through the Universe ever since the Universe became transparent
- **the relative abundance of hydrogen** and helium: stars and galaxies contain a ratio of about 7 to 1. This ratio is because the rest energy of a neutron is slightly greater than that of the proton.

The age of the Universe = $T \approx \frac{1}{H}$ (as the distance travelled by light in time T ($= cT$) ≈ maximum expansion distance $\frac{c}{H}$ given by Hubble's Law).

> **Revision tip**
> The tell-tale sign of a supernova is a sudden change of about 20 in its absolute magnitude over about 24 hours and then a gradual decrease over several years.

> **Synoptic link**
> Elements heavier than iron are formed by nuclear fusion in a supernova explosion. See Topic 27.3, Fission and fusion.

> **2.4 Summary questions**
>
> 1. **a** Describe the astronomical observations that indicate that a supernova event has occurred. *(3 marks)*
>   **b** Explain the preceding changes in a star that causes a supernova event. *(3 marks)*

> **3.1 Summary questions**
>
> $c = 3.0 \times 10^8 \text{ ms}^{-1}$
>
> 1. A spectral line of a certain star in a binary system changes from its laboratory wavelength of 620 nm by ± 0.072 nm with a time period of 2.5 years. Calculate the orbital speed of the star and the radius of its orbit. *(4 marks)*

Astrophysics

Dark energy

Distance measurements to type 1a supernovae by the red-shift method and the luminosity method give different results. This means that type 1a supernovae are much much further away than expected so they must be accelerating. Scientists think an unknown type of force is at work, releasing hidden energy referred to as **dark energy**.

At very large distances, dark energy is more prominent than gravity. Current theories suggest dark energy makes up about 70% of the total energy of the Universe.

Quasars

A quasar has a large red shift, indicating its distance is between 5000 and 10 000 light years away. Its light output is much greater than the Milky Way galaxy but its size is of the same order as the size of the Solar System.

Quasars are often found in or near distorted galaxies. These 'active' galaxies may have a **supermassive black hole** at their centres. Overheating of hot glowing gas as it is attracted towards the black hole results in the gas being thrown back out along its axis of rotation into space.

Summary questions

$c = 3.0 \times 10^8 \text{ ms}^{-1}$, $H = 65 \text{ kms}^{-1}\text{Mpc}^{-1}$

The absolute magnitude of any type 1a supernova is -18. A certain type 1a supernova has an apparent magnitude of $+22$.

1. **a** Calculate the distance to the supernova in parsecs. *(3 marks)*
   b Outline why measurements on type 1a supernovas have led to the conclusion that the expansion of the Universe is accelerating. *(2 marks)*

3.4 Exoplanets

Planets outside the Solar System are known as **exoplanets**.

- Many exoplanets are gas giants like Jupiter.
- Some exoplanets orbit within the habitable zone near a star (the zone in which liquid water may exist on a planet).

Methods of detection of exoplanets

1. **The radial velocity method:** a periodic Doppler shift is observed in the light from the star due to a planet and the star orbiting each other.
2. **The transit method:** the intensity of the light from the star regularly dips when an exoplanet passing 'in front' of the star blocks out some of the light from the star.

Summary questions

1. An exoplanet causes an intensity dip of 10% in the light intensity of a star. Estimate the radius of the exoplanet in terms of the radius of the parent star. *(3 marks)*

Medical physics

CHAPTER 1 PHYSICS OF THE EYE
1.1 Physics of vision

The ciliary muscle – concentric round the rim of the eye lens. To view a near object, the fibres contract to make the eye lens thicker. To view a distant object, the opposite happens. Adjustment is automatic and is called **accommodation**.

The iris – its concentric fibres contract and its radial fibres relax to make the eye pupil narrower so less light passes through it. In dim light, the opposite happens so more light passes through it.

The retina consists of rods and cones. Cones are colour sensitive and predominate at the fovea. Rods are sensitive to low intensity and are not colour sensitive.

To resolve two nearby point objects, their diffracted images on the retina need to be separated by at least two retinal cells.

The quantum efficiency of the eye is about 1 to 2% compared with ≈ 70% for a CCD.

1.2 Lenses
1.3 Defects of vision

Lenses

The power of a lens in dioptres is defined as $\dfrac{1}{\text{its focal length in metres}}$.

The lens formula $\dfrac{1}{u} + \dfrac{1}{v} = \dfrac{1}{f}$ where u is the distance from the lens to the object and v is the distance from the image to the lens.

Linear magnification = height of image / height of object = $\dfrac{v}{u}$.

Sight defects

The normal eye has a near point of 25 cm and a far point at infinity.

- **Myopia or short-sight** – cannot focus on distant objects.
- **Hypermetropia or long-sight** – cannot focus on nearby objects.
- **Astigmatism** – images sharper in focus in one direction than in other directions.

> **Revision tip**
> Make sure you can draw a graph to show the wavelength sensitivity of each type of cone and explain why a coloured after-image is in the complimentary colour.

> **Summary questions**
> 1 Sketch graphs on the same axes to show the response of the different types of retinal cells to light of different wavelengths from about 350 nm to 700 nm. Label each graph with the type of retinal cell. *(4 marks)*

> **Revision tip**
> Make sure you can draw ray diagrams to scale to locate an image formed by a converging lens and a diverging lens.

> **Revision tip**
> Make sure you can explain how sight defects are corrected, including calculation of the correcting lens focal length.

> **Summary questions**
> 1 a With the aid of a diagram, describe how myopia is corrected. *(4 marks)*
> b A short-sighted eye has a far point 3.00 m away. Calculate the power of the correcting lens to give a corrected far point at infinity. *(2 marks)*
> c A long-sighted eye has a near point at 0.50 m away. Calculate the power of the correcting lens to give a corrected near point at 25 cm from the eye. *(2 marks)*

Medical physics

> **Summary questions**
>
> 1. **a** State the functions of the bones of the middle ear. *(3 marks)*
>   **b** Explain why the pressure variations on the oval window from the middle ear are much greater than the pressure variations of sound waves on the ear drum. *(3 marks)*

> **Revision tip**
> Make sure you can draw a graph to show how the least detectable intensity level of a normal ear varies with frequency.

CHAPTER 2 PHYSICS OF THE EAR

2.1 The structure of the ear

The normal human ear has a frequency range from about 20 Hz to 18 000 Hz.

The middle ear – the three bones: the hammer, the anvil, and the stirrup, transmit the vibrations of the tympanic membrane to the oval window of the inner ear. They:

- increase the force of the vibrations on the oval window
- filter out unwanted noise
- protect the ear from high-sound levels.

The inner ear – receives sound vibrations from the middle ear at its **oval window**, causing vibrations in the **cochlea** fluid which make the **basilar membrane** vibrate. Tiny sensitive 'hair cells' on the membrane then send nerve impulses to the brain.

2.2 Sound measurements
2.3 Frequency response

The decibel scale

The intensity of a sound wave (and of any type of wave) is the energy per second per unit area incident normally on a surface. The unit of intensity is $Js^{-1}m^{-2}$ or Wm^{-2}.

The threshold intensity of hearing I_0 is the least intensity the human ear can hear at a frequency of 1000 Hz (= 1 picowatt / m²).

Intensity level in decibels dB = $10\log\dfrac{I}{I_0}$

The dBA scale matches the normal ear's response at different frequencies.

Hearing loss in dB at any given frequency is the difference between the threshold intensity level for normal hearing and for the person being tested.

> **Summary questions**
>
> $I_0 = 1.0 \times 10^{-12}\,Wm^{-2}$
>
> 1. **a** A machine creates sound at an intensity level of 115 dB at a certain distance from it. Calculate the intensity of the sound waves at this distance. *(3 marks)*
>   **b** Two outdoor loudspeakers each create an intensity level of 95 dB at a certain position. Calculate the intensity level at this position. *(3 marks)*

CHAPTER 3 BIOLOGICAL MEASUREMENT

3.1 ECG signals

Figure 1 shows a typical ECG trace (or electrocardiogram) for a healthy person.

The main features of the trace are labelled as P, Q, R, S, and T.

- **P** is when the atria contract. Between P and Q, the ventricles fill.
- **QRS** is when the ventricles contract and the atria relax.
- **T** is when the ventricles relax.

An ECG amplifier needs an even voltage gain of ≈ 1000 up to about 20 Hz, a high signal-to-noise ratio, and a high input impedance.

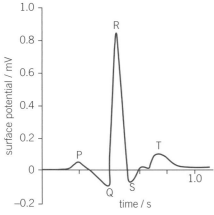

▲ **Figure 1** *An ECG trace of a healthy person*

Medical physics

Summary questions

1. **a i** Draw a labelled diagram of an ECG trace obtained from a person with a healthy heart, showing how the pd varies with time during one heartbeat. *(2 marks)*
   ii Label the main features on your trace and describe how they relate to the changes in the heart that take place during one heartbeat. *(5 marks)*
   b State two essential characteristics of an ECG amplifier other than high input impedance, giving a reason why each characteristic is essential. *(4 marks)*

CHAPTER 4 NON-IONISING IMAGING

4.1 Ultrasonic imaging

In an **A-scan system**, an oscilloscope is used to measure the time taken t by the reflected pulse to travel from the probe to the reflecting boundary and back. The distance, x, from the body surface to the reflecting boundary $= \frac{1}{2}ct$, where c is the speed of ultrasonic waves in the body.

In an **B-scan system**, the probe contains several transducers which transmit pulses simultaneously. As the probe is moved across the patient, the B-scan system therefore gives a two-dimensional image.

Reflection of ultrasonic waves

The **acoustic impedance** $Z = \rho c$ where ρ is the density of the substance and c is the speed of ultrasonic waves in it. The unit of acoustic impedance is $kg\,m^{-2}\,s^{-1}$.

- The intensity of waves reflected at a boundary, $I_R = \dfrac{(Z_2 - Z_1)^2}{(Z_2 + Z_1)^2} \times I_0$ where I_0 is the incident intensity and where Z_1 and Z_2 are the acoustic impedances of the incident and reflected intensities respectively.

- A gel is applied to the skin where the probe is applied to prevent reflection of ultrasonic waves at the body surface.

4.2 Endoscopy

The principle of the endoscope

The **incoherent** bundle of an endoscope transmits light into the body. A lens over the end of the **coherent bundle** is used to form an image of the body cavity on the end of the fibre bundle. The fibre ends are in the same relative positions at each end so the image formed by the lens inside the body is seen as a 'coherent' image on the end of the bundle outside the body.

The effect of different refractive index values

The angle of the viewing cone $= 2i_{max}$ where i_{max} is given by $n_0 \sin i_{max} = n_1 \sin(90 - i_c)$ where n_0 and n_1 = refractive indices of air and the core respectively.

For no light loss from a fibre of diameter d, the maximum radius of curvature $R = \dfrac{d}{\dfrac{n_1}{n_2} - 1}$

where n_2 = refractive index of the cladding.

Summary questions

1. **a** Explain the functions of the backing block in an ultrasonic probe. *(2 marks)*
   b Explain why a gel must be applied between an ultrasonic probe and the skin when the probe is used. *(2 marks)*
   c An organ has a density of $1070\,kg\,m^{-3}$ and the speed of sound through it is $1580\,m\,s^{-1}$. The corresponding data for surrounding soft tissue is $1050\,kg\,m^{-3}$ and $1550\,m\,s^{-1}$. Calculate the reflection coefficient of the boundary between the organ and the surrounding soft tissue. *(2 marks)*

Synoptic link

See Topic 5.3, Total internal reflection, to remind yourself about the law of refraction, total internal reflection, and optical fibres.

Medical physics

Summary questions

1. **a** State the function of a coherent bundle and of an incoherent bundle in an endoscope. *(3 marks)*
   b Explain why the optical fibres in the coherent bundle of an endoscope need to be very thin and why they should not be bent too much. *(2 marks)*
   c Calculate the angle of the viewing cone for a fibre which has a core of refractive index 1.60 and a refractive index 1.52. *(4 marks)*

4.3 The MR scanner

A hydrogen nucleus precesses in a magnetic field at two different energy levels. The frequency of precession depends on the magnetic flux density. When a pulse of radio waves is applied, at any point where the precession frequency is equal to the radio wave frequency:

- nuclear magnetic resonance occurs and nuclei in the lower energy state flip into the higher energy level
- excited nuclei flip back to the lower energy state, each emitting a radio wave photon.

In the MR scanner

In a scan, the magnetic field is designed to change the location of the excited nuclei systematically The signal detected after each pulse is from that location. The detector is connected to a computer which creates a visual image of a cross-section of the patient.

Tissue discrimination is possible because the rate of decay of the signal, the relaxation time, depends on the type of molecules surrounding the hydrogen molecules.

MRI scans are non-ionising radiation unlike X-ray scans.

Summary questions

1. **a** Explain how an MR scanner is used to detect the location of hydrogen nuclei in the body. *(5 marks)*
   b What is meant by relaxation time in the context of an MR scan and why is it an important consideration in an MR scan? *(3 marks)*

CHAPTER 5 X-RAY IMAGING

5.1 The physics of diagnostic X-rays

The anode pd controls the maximum energy of the X-ray photons from an X-ray tube. The minimum wavelength $\lambda_{MIN} = \dfrac{hc}{eV}$.

The electron beam current controls the intensity of the X-ray beam.

The electrical power supplied to an X-ray tube = beam current $I \times$ anode pd V.

> **Revision tip**
> Make sure you know how an X-ray tube works and that you can draw a graph to show how the relative intensity of the X-rays varies with wavelength for a certain tube voltage V.

Summary questions

$e = 1.6 \times 10^{-19}\,\text{C}, h = 6.63 \times 10^{-34}\,\text{Js}, c = 3.0 \times 10^{-8}\,\text{ms}^{-1}$

1. **a** Sketch a graph to show how the intensity of the X-rays from an X-ray tube varies with wavelength. *(3 marks)*
   b An X-ray tube operates at a pd of 60 kV. Calculate the minimum wavelength of the X-rays emitted. *(2 marks)*

5.2 X-rays and matter

When a beam of X-rays passes through an 'absorber' of thickness x, the transmitted intensity I is given by

$$I = I_0 e^{-\mu x}$$

where I_0 is the incident intensity of the beam and μ is the **attenuation coefficient** (sometimes called the absorption coefficient) of the absorber substance. The unit of μ is metre^{-1}.

The half-value thickness, $X_{\frac{1}{2}}$, of an absorber of X-rays is defined as the thickness required to reduce the intensity by 50%. Note that $X_{\frac{1}{2}} = \dfrac{\ln 2}{\mu}$

The mass attenuation coefficient of a substance $\mu_m = \dfrac{\mu}{\rho}$. The unit of μ_m is m² kg^{-1}.

> **Revision tip**
> Make sure you can explain in terms of μ_m why a metal filter is used to remove low energy photons (<30 keV) from the beam. Also, why photons of energy >100 keV are not used for radioagraphy.

Summary questions

1. Bone has a density of 1900 kg m^{-3} and a half-value thickness for 50 keV X-ray photons of 15 mm.
   a. Calculate the mass attenuation coefficient of bone. *(3 marks)*
   b. Calculate the % reduction in intensity when 50 keV X-rays pass through 4.0 mm of bone. *(3 marks)*

5.3 Image enhancement

Image quality

The image sharpness is improved by making the anode surface at an angle of about 70° to the electron beam so that the X-rays effectively originate from a much smaller area than the impact area of the beam.

The image contrast is improved by using:

- **a lead collimator grid** to prevent X-rays scattered in the patient from reaching the film
- **a contrast medium** (e.g., barium) as it is a good absorber of X-rays and can make the image of certain organs stand out.

Image intensification

The intensifying screen – a double-sided film between two sheets of fluorescent material that absorb X-rays photons and re-emit many more light photons.

The image intensifier allows the radiation dose to be reduced by a factor of 1000 and enables a 'real time' image to be observed and recorded.

A flat panel (FTP) detector used in place of a photographic film gives a digital image on a separate LCD screen.

The CT scanner

A CT (computed tomography) scanner consists of an X-ray tube and a ring of thousands of small solid-state detectors linked to a computer. The tube moves round the ring with the beam always directed at the patient at the centre of the ring. The detector signals are processed to display a 2D digital image of the cross-section of the patient.

An X-ray tube in a CT scanner operates at potentials between 100 and 150 kV and the X-rays are filtered so they are monochromatic (i.e., monoenergetic).

> **Revision tip**
> Ensure you can explain how an image intensifier works.

> **Revision tip**
> Compare an FTP with photographic film in terms of sensitivity, image distortion, response speed, and 'digital' benefits.

Summary questions

1. a. State and explain two methods used to improve the contrast of an X-ray image. *(6 marks)*
   b. State the purpose of an intensifying screen and explain how its design achieves this purpose. *(4 marks)*

> **Synoptic link**
> Te-99m can be prepared 'on-site' in a **technetium generator**. See Topic 26.8, More about decay modes.

Medical physics

> **Revision tip**
> Make sure you can explain how a photomultiplier works.

Summary questions

1. **a** A radioactive isotope has a physical half-life of 12.0 hours and a biological half-life of 24 hours. Calculate the effective half-life of this isotope in the body. *(2 marks)*
   b The technetium isotope $^{99}_{43}Tc$ is metastable and has a half-life of 6.0 hours. State two reasons why this isotope is often used for diagnostic purposes. *(2 marks)*

Summary questions

1. **a** Give two reasons why X-rays from a 100 kV X-ray tube are less suitable for medical therapy compared with X-ray radiation from an electron accelerator. *(2 marks)*
   b The lutecium $^{177}_{71}Lu$ isotope, a β^- emitter with a half-life of 6.7 days, given intravenously, attaches itself to certain cancer cells and is used in the treatment of tumours that have spread into different parts of the body. The isotope emits 0.5 MeV β^- particles. Explain why 0.5 MeV X-ray photons from an electron accelerator could not be suitable for this purpose. *(2 marks)*

CHAPTER 6 RADIONUCLIDE IMAGING AND THERAPY

6.1 Radionuclide imaging

Properties of radioisotopes

The **effective half-life** T_{eff} of a radioactive isotope in the body is the time taken for the activity of the radioactive isotope in the body to reduce to half its initial activity.

$$\frac{1}{T_{eff}} = \frac{1}{T_b} + \frac{1}{T_p}$$

where T_p is **the physical half-life** of the radioactive isotope and T_b is the **biological half-life** of the substance in the body.

Most radioisotopes are produced by exposing stable isotopes to neutron radiation in a nuclear reactor or from a particle accelerator so the exposed isotope becomes unstable and emits γ radiation after β emission. The technetium isotope $^{99}_{43}Tc$ is used for many diagnostic purposes as it forms in a 'metastable' state, usually referred to as Tc-99m, which then decays with a short half-life of 6 hours by emitting β radiation with a much longer half-life.

A **gamma camera** is used to detect the γ photons from the radioisotopes. It uses **photomultipliers** in which each γ photon incident on the crystal of the gamma camera generates an electronic pulse.

The PET (positron emission tomography) scanner

In a PET scan:

- Each positron travels less than a millimetre because it meets an electron and they annihilate each other, producing two γ photons travelling in opposite directions.

- A ring of detectors connected to a computer registers a positron emission when opposite detectors detect a γ photon at the same time.

6.2 Radiation therapy

X-ray tubes for therapy do not operate above about 300 kV because their insulation breaks down at such high potential differences. Their X-rays are not penetrating enough to reach tumours deep inside the body.

Electron accelerators can generate X-ray beams with energies up to 25 MeV which are capable of destroying deep tumours. Electron accelerators can be switched on and off as required, enabling exposure times to be carefully controlled. In addition, they do not contain highly-radioactive sources and extensive shielding is not required round the source of radiation (i.e., the target).

Radioactive implants are used to destroy malignant cells in a tumour. Radioactive isotopes used include:

- isotopes that emit monoenergetic gamma photons so they have little effect beyond their intended range
- beta-emitting isotopes with half-lives of the order of days.

Engineering physics

CHAPTER 1 ROTATIONAL DYNAMICS

1.1 Angular acceleration

For constant angular acceleration of a rotating body. If the change of angular velocity is $\Delta\omega$ in a time interval Δt

$$\text{its angular acceleration, } \alpha = \frac{\Delta\omega}{\Delta t}$$

The unit of angular displacement is the radian per second (rad s^{-1}) where 2π radians = 360°. In a rotating body, for a point (i.e., small 'element') in the body at perpendicular distance r from the axis of rotation:

- its speed $v = \omega r$
- its tangential acceleration $a = \alpha r$.

Equations for constant angular acceleration

For any body rotating with a constant angular acceleration α about a fixed axis with an initial angular speed ω_1, and an angular speed ω_2, and angular displacement θ time t later:

- $\omega_2 = \omega_1 + \alpha t$
- $\theta = \frac{1}{2}(\omega_2 + \omega_1)t$
- $\theta = \omega_1 t + \frac{1}{2}\alpha t^2$
- $\omega_2^2 = \omega_1^2 + 2\alpha\theta$

The average value of the angular speed $= \frac{1}{2}(\omega_2 + \omega_1)$

Angular motion graphs

Angular motion graph	Equivalent linear motion graph
1 **Angular displacement θ against time t**	1 **Displacement s against time t**
Gradient = angular velocity ω	Gradient = velocity v
2 **Angular velocity ω against time t**	2 **Velocity v against time t**
Gradient = angular acceleration α	Gradient = (linear) acceleration a
Area under the graph line = angular displacement	Area under the graph line = displacement s

Summary questions

1 A flywheel accelerates from 2.0 to 17.0 revolutions per minute in 75.0 s. The radius of the flywheel is 0.036 m. Calculate:
   a the angular acceleration of the flywheel *(2 marks)*
   b the number of turns it makes. *(3 marks)*

1.2 Moment of inertia

Torque $T = F r$ where the force F that causes the torque acts at perpendicular distance r from the axis to the line of action of the force. The unit of torque is the newton metre (Nm).

The moment of inertia I of a body about a given axis is defined as $\Sigma m_i r_i^2$ for all the points in the body, where m_i is the mass of each point and r_i is its perpendicular distance from the axis. The unit of I is kgm^2.

When a body undergoes angular acceleration α, the resultant torque T acting on it is given by $T = I\alpha$. Two bodies of equal mass distributed in different ways will have different values of I.

In general, **the further the mass is distributed from the axis, the greater is the moment of inertia about that axis.**

Synoptic link

See Topic 8.3, Terminal speed, for the linear dynamics equations with constant acceleration

Revision tip

For a circular hoop of mass M and radius R, its moment of inertia about an axis through the centre perpendicular to its plane = MR^2.

Summary questions

1 A flywheel is accelerated by a constant torque for 38.2 s from rest. During this time it makes 24.6 turns. It then slows down to a standstill 92.0 s after the torque is removed, making 97.0 turns during this time.
   a i Calculate its average angular acceleration when it was being accelerated. *(3 marks)*
   ii Calculate its average angular deceleration when it was slowing down. *(2 marks)*
   b The moment of inertia of the flywheel was 290 kgm^2. Calculate the torque applied to it when it was being accelerated. *(3 marks)*

Engineering physics

Summary questions

1. A 1.85 kg object hanging from a string wrapped round the axle of a flywheel of diameter 12.0 mm is used to accelerate the flywheel which is on frictionless bearings from rest. The object falls through a vertical distance of 2.2 m in 9.4 s which is the time the string takes to unwrap from the axle. Calculate:
   a the potential energy lost by the object in descending 2.2 m *(1 mark)*
   b the kinetic energy of the object 9.4 s after it was released from rest *Hint: Assume the acceleration of falling object is constant and calculate its speed from its initial speed, the distance it moved and the time it took.* *(2 marks)*
   c i the kinetic energy gained by the flywheel *(1 mark)*
     ii the moment of inertia of the flywheel. *(3 marks)*

1.3 Rotational kinetic energy

A body rotating at angular speed ω has kinetic energy $E_K = \frac{1}{2}I\omega^2$ where I is its moment of inertia about the axis of rotation. **The work done W by a constant torque when the body is turned through angle $\theta = T\theta$.** For a body rotating at constant angular speed ω, the **power P** delivered by a torque T acting on the body is given by

$$P = T\omega$$

Uses of flywheels

1. To smooth out the variations of the speed of a machine or an engine when the load varies.
2. To store kinetic energy when a vehicle's brakes are applied and it slows down.

1.4 Angular momentum

The angular momentum of a rotating object $= I\omega$ where I is the moment of inertia of the body about the axis of rotation and ω is its angular speed. The unit of angular momentum is $kg\,m^2\,rad\,s^{-1}$ or $N\,m\,s$.

Conservation of angular momentum

The total angular momentum of any system consisting of one or more rotating objects is constant, provided the resultant torque on the system is zero. **The angular impulse of a short-duration torque T is defined as $T\,\Delta t$.** If $\Delta\omega$ is the change of angular momentum due to this torque T, then **angular impulse $T\Delta t$ = change of angular momentum $I\Delta\omega$**

▼ **Table 1** *Comparison between linear and rotational motion*

Linear motion	Rotational motion
displacement s	angular displacement θ
speed and velocity v	angular speed ω
acceleration a	angular acceleration α
mass m	moment of inertia I
momentum mv	angular momentum $I\omega$
force F $F = ma$	torque $T = Fd$ $T = I\omega$
$F = \frac{d(mv)}{dt}$	$T = \frac{d(I\omega)}{dt}$
Impulse $F\Delta t = m\Delta v$	Angular impulse $T\Delta t = I\Delta\omega$
kinetic energy $= \frac{1}{2}mv^2$	kinetic energy $= \frac{1}{2}I\omega^2$
work done $= Fs$	work done $= T\theta$
power $= Fv$	power $= T\omega$

Summary questions

1. A metal disc X on the end of an axle rotates freely at 180 revolutions per minute. The moment of inertia of the disc is $0.034\,kg\,m^2$.
   a Calculate the angular momentum of the disc. *(2 marks)*
   b After a second disc Y that is initially stationary on the same axis is engaged by X, both discs rotate at 125 revolutions per second. Calculate the moment of inertia of Y. *(4 marks)*

Engineering physics

CHAPTER 2 THERMODYNAMICS
2.1 The first law of thermodynamics
2.2 Thermodynamics of ideal gases

The first law of thermodynamics states that heat energy (energy transferred by heating) supplied to a system either increases the internal energy of the system or enables it to do work, or both. If the internal energy of the object increases by ΔU, and the work done **by** the system is W,

then the heat transfer into the system, $Q = \Delta U + W$

▲ **Figure 1** *Change of internal energy*

A negative value for ΔU means the internal energy has decreased. A negative value for Q means heat energy is transferred out of the system. A negative value for W means that work is done **on** the system.

Work done by an ideal gas
- Work is done on a gas when its volume is reduced (i.e., it is compressed).
- Work is done by a gas when it expands.
- Consider a tube of gas at pressure p trapped by a piston, as in Figure 2. When the gas expands,

work done by the gas, $W = p\Delta V$

Adiabatic changes of an ideal gas

For any adiabatic change of an ideal gas, its pressure p and volume V are given by the equation

$$pV^\gamma = \text{constant}$$

where γ is the adiabatic constant of the gas.

1. You will be given the value of γ in an examination question.
2. Adiabatic changes of an ideal gas can be plotted on a graph of pressure against volume. The curves are steeper than isotherms (which are constant temperature curves in accordance with pV = constant), as Figure 3 shows.

Pressure against volume curves

The area under any pV curve represents the work done on or by the gas.

> **Synoptic link**
> See Topic 11.1, Density, for internal energy.

▲ **Figure 2** *Work done by a gas*

> **Revision tip**
> An energy transfer where no heat transfer occurs is called an **adiabatic change**.
>
> An energy transfer where the temperature is constant is called an **isothermal change**.

> **Synoptic link**
> See Topic 20.2, The ideal gas law, to remind yourself about the properties of an ideal gas.

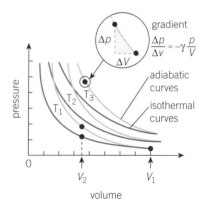

▲ **Figure 3** *Adiabatic curves*

Summary questions

1. **a** When a gas expands, it gains 100 J of internal energy and does 80 J of work. Calculate the heat transfer to or from the gas. *(1 mark)*
   b When a gas is compressed and 80 J of heat energy is transferred from it, it gains 100 J of internal energy. Calculate the work done on or by the gas. *(1 mark)*

2. **a** A cylinder contains 0.0420 m³ of air at a pressure of 153 kPa and a temperature of 320 K. The air in the cylinder expands adiabatically to a volume of 0.0542 m³. Calculate the pressure and the temperature of the air immediately after it has been compressed. $\gamma = 1.40$ for air *(5 marks)*
   b Estimate the work done by the gas. *(3 marks)*

Engineering physics

> **Revision tip**
> No engine can have an efficiency of 100% or more.

2.3 Heat engines

A **heat engine** is designed to do useful work which is supplied to it as a result of heat transfer. For an engine to do work as a result of heat transfer from a high temperature source, we need a low temperature sink to draw the energy from the source. Assuming $\Delta U = 0$ for the engine,

$$Q_H = W + Q_C$$

where Q_H = heat transfer to the engine, W = work done by the engine, Q_C = heat losses to the surroundings.

The second law of thermodynamics states that it is impossible for heat transfer from a high temperature source to produce an equal amount of work.

The efficiency of an engine is defined as

$$\frac{\text{the work done by the engine}}{\text{energy supplied to the engine}}$$

Engine cycles
The petrol engine

The petrol engine uses air as the working substance, heating it by internal combustion. The indicator diagram in Figure 1 shows the pressure changes in one cylinder of a four-cylinder petrol engine in one complete cycle. The cylinders 'fire' sequentially to maintain the motion of each cylinder.

▲ **Figure 1** *The four-stroke petrol engine*

A → B Adiabatic compression No heat enters or leaves the cylinder. The work done on the air increases its internal energy.

B → C Ignition A spark ignites the mixture, heating the gas and raising its pressure and temperature in the cylinder very rapidly. Heat Q_H is gained by the gas during ignition.

C → D Power Work is done adiabatically by the expanding gas.

D → A Cooling The gas is cooled rapidly by contact with the internal walls of the cylinder block, reducing the pressure to the initial pressure. Heat Q_C is lost from the gas during cooling.

> **Revision tip**
> Friction power = indicated power − brake power.

2.3 Summary questions

1. A vehicle engine with 4 cylinders uses fuel of calorific value 45 MJkg^{-1} at a rate of 2.9×10^{-3} kgs^{-1} when it operates at 37 cycles per second. In this situation, the work done by each cylinder in each engine cycle is 0.28 kW. Calculate:
   a. the input power from the engine *(2 marks)*
   b. the thermal efficiency of the engine. *(3 marks)*

The net work done W by one engine cylinder in one cycle = $Q_H - Q_C$ = the area of the indicator loop ABCD. The **indicated power** of the engine = the rate of net work done by the engine per second = area of indicator loop × number of cycles per second × number of cylinders.

The **input power** to the engine = the calorific value of the fuel (in J kg⁻¹) × the fuel flow rate (in kgs⁻¹) where the **calorific value** of the fuel is the energy released per unit mass of fuel when the fuel is burnt.

The thermal efficiency of the engine = $\dfrac{\text{work done per second}}{\text{energy supplied per second}}$

$= \dfrac{\text{indicated power}}{\text{input power from fuel combustion}}$

- The mechanical efficiency of an engine

$= \dfrac{\text{brake power } T\omega \text{ (i.e., output power)}}{\text{indicated power}}$

- The overall efficiency = its mechanical efficiency × its thermal efficiency

2.4 Efficiency and thermodynamics

For any heat engine operating between a heat source and a heat sink:

- Its % efficiency of = $\dfrac{W}{Q_H} \times 100\% = \dfrac{Q_H - Q_C}{Q_H} \times 100\%$ (since $W = Q_H - Q_C$).
- The most efficient type of engine is a theoretical engine called **a reversible engine**.
- For a reversible engine, $\dfrac{T_C}{T_H} = \dfrac{Q_C}{Q_H}$.

The % efficiency of a reversible engine = $\dfrac{T_H - T_C}{T_H} \times 100\%$.

Efficiency in practice

Thermal efficiency is less than above because:

- the indicator loop area in practice does not have sharp corners as in the theoretical diagram so less work is done (due to the valves not opening and closing instantly)
- the maximum temperature is not reached as in perfect combustion
- the expansion and compression strokes are not perfectly adiabatic as assumed in calculating the theoretical thermal efficiency
- work is done by the engine to bring about the induction and exhaust strokes.

Mechanical efficiency is less than 100% because:

- friction between the moving parts in the engine cannot be eliminated
- oil used to lubricate the bearings in an engine are viscous which means they cause some resistance to the motion of the moving parts.

2.5 Heat pumps and refrigerators

A heat pump is a device that transfers energy from a cold space to a hot space.

Where work done W causes heat transfer Q_C from the cold space and heat transfer Q_H (= $Q_C + W$) into the hot space.

its coefficient of performance = $\dfrac{Q_H}{W} = \dfrac{Q_H}{Q_H - Q_C}$

If T_H and T_C are the temperatures of the hot 'space' and the cold 'space' respectively, its coefficient of performance $< \dfrac{T_H}{T_H - T_C}$

In a refrigerator, work is done in order to remove energy from food and drink in a cold space. The coefficient of performance of a refrigerator is defined as $\dfrac{Q_C}{W}$ (in comparison with $\dfrac{Q_H}{W}$ for a heat pump). Therefore,

its coefficient of performance = $\dfrac{Q_C}{W} = \dfrac{Q_C}{Q_H - Q_C}$

If T_H and T_C are the temperatures of the hot 'space' and the cold 'space' respectively, its coefficient of performance $< \dfrac{T_C}{T_H - T_C}$

2.4 Summary questions

1. A gas-powered engine operates between temperatures of 1100 K and 300 K. When the engine uses gas of calorific value 31 MJ kg⁻¹ supplied at a rate of 9.4 kg h⁻¹, the mechanical power output of the engine is 38 kW.
   a. Calculate the theoretical efficiency of the engine. *(2 marks)*
   b. i. Show that the input power to the engine is 81 kW. *(2 marks)*
   ii. Calculate the heat transferred by the heat engine to its low-temperature sink. *(1 mark)*

Summary questions

1. A heat pump fitted with a 120 W electric motor supplies 600 W of heat per second to a room from outdoors. Calculate:
   a. its coefficient of performance *(2 marks)*
   b. the heat gained from outdoors. *(1 mark)*

2. A refrigerator with a 5 W pump cools a plastic beaker containing 0.18 kg of water from 20°C to 0°C in 1200 s. Calculate:
   a. the heat energy extracted from the water *(2 marks)*
   b. the coefficient of performance of the refrigerator.
   specific heat capacity of water = 4200 J kg⁻¹ K⁻¹ *(2 marks)*

Turning points in physics

Summary questions

1. **a** Explain why the positive ions in a discharge tube depend on the gas in the tube but the negative particles are identical. *(4 marks)*

   b Electrons in a vacuum tube reach a speed of 3.6×10^7 ms^{-1} after being accelerated from rest through a potential difference V. Calculate the speed of the electrons if they are accelerated through a potential difference $\frac{1}{2}V$. *(2 marks)*

Synoptic link

The speed of the electrons can be determined using the perpendicular E and B fields as in the velocity selector in Topic 24.2, Moving charges in a magnetic field.

Summary questions

1. Electrons moving at a constant speed are directed horizontally into a uniform electric field due to two parallel plates spaced 40.0 mm apart which have a pd of 5000 V between them.

   a When a uniform magnetic field of flux density 3.25 mT is applied at right angles to the beam and to the electric field, the beam is undeflected. Calculate the speed of the electrons. *(2 marks)*

CHAPTER 1 THE DISCOVERY OF THE ELECTRON

1.1 Thermionic emission of electrons

Cathode rays

When a sufficiently large pd is applied to a discharge tube, the gas conducts and emits light.

- Positive ions and electrons produced by ionisation near the cathode recombine and emit light photons.
- Electrons that do not recombine (called cathode rays) accelerate and move towards the anode. They excite gas atoms by collision further along the tube. These atoms de-excite and emit light photons.

The principle of thermionic emission

Electrons from a heated filament wire in a vacuum tube are attracted to the anode where some pass through a small hole to form a narrow beam. The speed v of each electron in the beam is given by $\frac{1}{2}mv^2 = eV$.

1.2 Deflection of an electron beam
1.3 Use of electric and magnetic fields to determine $\frac{e}{m}$

Using a uniform magnetic field

Electrons of speed v are deflected in a circular path. The centripetal force on each electron ($= \frac{mv^2}{r}$) is due to the magnetic force ($= Bev$) hence $r = \frac{mv}{Be}$.

Combining this equation with the anode pd equation $\frac{1}{2}mv^2 = eV_A$ gives $\frac{e}{m} = \frac{2V_A}{B^2 r^2}$.

Using graphs

A graph of r against $\frac{1}{B}$ gives a straight line through the origin with a gradient $k = \left(\frac{2mV_A}{e}\right)^{\frac{1}{2}}$.

The value of k is used to determine $\frac{e}{m}$.

Using a uniform electric field

The electric field is between two parallel plates of length L and spacing d at pd V_p, electrons of speed v are directed into the field perpendicular to the field and the deflection y of the beam is measured.

- The acceleration, a towards the positive plate $= \frac{2y}{t^2}$ (from $y = \frac{1}{2}at^2$) where $t = \frac{L}{v}$.
- The force on each electron, $F = \frac{eV_p}{d}$.

Therefore, the acceleration $a = \frac{F}{m} = \frac{eV_p}{md}$ so $\frac{e}{m} = \frac{ad}{V_p}$.

The significance of Thomson's determination of $\frac{e}{m}$

JJ Thomson showed that the electron's specific charge $\frac{e}{m}$ was 1860 × larger than that of the hydrogen ion which had the largest known specific charge. But neither the mass nor the charge of the electron was known at that time.

1.4 The determination of the charge of the electron, e, by Millikan's method

For a droplet of charge Q and mass m between oppositely charged horizontal parallel metal plates, when the droplet is stationary, $\frac{QV_p}{d} = mg$. Hence $Q = \frac{mgd}{V_p}$.

To measure the droplet's mass, the electric field is switched off and the droplet's terminal speed is measured. At this speed, the droplet weight ($\frac{4}{3}\pi r^3 \rho g$) = the drag force on it ($6\pi \eta r v$) where η is the viscosity of air. Therefore $r^2 = \frac{9\eta v}{2\rho g}$ is used to calculate the droplet radius r and its mass.

The significance of Millikan's results

Millikan discovered that the electric charge Q is **quantised** in whole number multiples of 1.6×10^{-19} C. He therefore concluded that the charge of the electron is -1.6×10^{-19} C.

Summary questions

$g = 9.81$ ms^{-2}, $e = 1.6 \times 10^{-19}$ C

1. A charged oil droplet was held stationary between oppositely charged metal plates 5.50 mm apart when the pd between the plates was 696 V. When the pd was switched off, the droplet fell at a terminal speed of 1.55×10^{-4} ms^{-1}. Calculate:
   a. the mass of the droplet *(3 marks)*
   b. the charge of the droplet. *(2 marks)*
   Density of the oil = 960 kg m^{-3}, viscosity of air = 1.80×10^{-5} Nsm^{-2}

CHAPTER 2 WAVE PARTICLE DUALITY

2.1 Early theories of light

Newton's corpuscular theory of light	• Light consists of 'corpuscles' (i.e., tiny particles).	• reflection explained by assuming corpuscles bounce off a mirror without loss of speed. • refraction explained by assuming corpuscles travel faster in a refractive substance than in air.
Huygens' wave theory of light	• A wavefront propagates by creating wavelets that move forwards and recreate the wavefront.	• reflection explained by assuming wavefronts reflect off a mirror without loss of speed and the speed of the wavelets is unchanged. • refraction explained by assuming the waves travel **slower** in a transparent substance than in air.
Wave theory was rejected in favour of corpuscular theory because: • *there was no experimental evidence to show light travels faster or slower in a transparent substance than in air* • *Newton had a much stronger scientific reputation than Huygens* • *light waves considered to be longitudinal so could not explain polarisation of light.*		
Young's double slits experiment	• Light passed through double slits produces an interference pattern.	• Corpuscular theory of light incorrectly predicted there would be only two bright fringes. Wave theory can explain interference.
The wave theory of light was not accepted for many years until it was shown that light in water travels slower than in air.		

Turning points in physics

b. When the electric field is switched off, the beam is deflected to form a circle of diameter 68.0 mm. Calculate the specific charge, $\frac{e}{m}$, of the electron. *(2 marks)*

Revision tip
Make sure you can explain Newton's ideas on reflection and refraction in terms of velocity components.

Synoptic links
Look back at Topic 5.1, Refraction of light, and Topic 5.4, Double slit interference.

Summary questions

1. a. Explain how Newton's corpuscular theory of light explains the refraction of a light ray when it passes from glass into air. *(3 marks)*
   b. Explain why Newton's theory of light cannot explain Young's fringes pattern whereas Huygens' wave theory can. *(4 marks)*

Turning points in physics

Summary questions

1 a Hertz discovered that the strength of the radio signal detected from a radio wave transmitter aerial varies according to the orientation of the detector. Explain this effect and state the conclusion drawn by Hertz about the radio waves from the transmitter. *(3 marks)*

b In an experiment to measure the speed of light by Fizeau's method, the reflected light could not be observed when the cog wheel with 720 teeth was rotating at 12.0 Hz and the distance travelled by the light was 8.5 km. Use this data to calculate the speed of light in air. *(2 marks)*

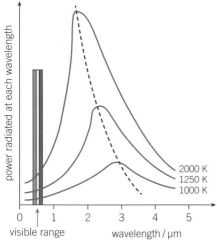

▲ **Figure 1** *Black-body radiation curves*

2.2 The discovery of electromagnetic waves

Maxwell's theory Electromagnetic waves are transverse electric and magnetic waves that are in phase and perpendicular to each other and the direction of propagation.	Maxwell derived $c = \dfrac{1}{\sqrt{\mu_0 \varepsilon_0}}$ for their speed in free space and showed it is the same as the speed of light.

Hertz's discovery
• A spark gap transmitter was used to produce radio waves and a metal loop to detect them. The waves caused sparks to jump across the gap in a detector loop.
• Hertz also showed that radio waves
  • can be polarised
  • are reflected by a metal sheet and can form stationary waves.
	• Hertz measured the distance between adjacent nodes ($= 0.5\lambda$) of stationary radio waves of known frequency and used $v = f\lambda$ to find their speed.
	
	 Note that $t = 1/(2 f_0 N)$ as there are $2N$ teeth and gaps round the cogwheel perimeter and the cogwheel's period of rotation is $1/f_0$.

Fizeau's measurement of the speed of light A rotating cog wheel was used to chop a narrow beam of light into pulses of light. The pulses were reflected back by a distant reflector. As the rotation frequency of the cog wheel was increased, the reflected light was blocked at a certain frequency f_0, corresponding to the time taken, t, for a gap to be replaced by the next tooth.	$c = \dfrac{2D}{t} = 4Df_0 N$ where D is the distance from the cog wheel to the reflector and N is the number of teeth on the cogwheel.

2.3 The development of the photon theory of light

The ultraviolet catastrophe

Figure 1 shows how the intensity of the radiation from a 'black body' (i.e., a perfect absorber) varies with wavelength for different temperatures of the object. Wave theory predicts incorrectly that the intensity becomes infinite at smaller and smaller wavelengths. Planck solved the problem by introducing the idea that the energy of an atom vibrating at frequency f can only be an integral multiple of hf.

The discovery of photoelectricity

Experiments showed that for any given metal:

- no photoelectrons are emitted if the frequency of the incident light is below a threshold frequency
- photoelectric emission occurs instantly.

Wave theory predicts incorrectly that light of any frequency causes photoelectric emission and the lower the frequency of the light, the longer photoelectric emission takes.

Einstein's photon theory of light states that light consists of photons, each carrying energy hf, where f is the frequency of the radiation.

For a conduction electron to escape, Einstein said it needs to:

1. absorb a single photon of energy hf
2. use energy equal to the work function ϕ of the metal to escape.

Hence, the **threshold frequency** of the incident radiation $f_0 = \dfrac{\phi}{h}$ as $hf \geq \phi$.

Using graphs

Photoelectric emission from a metal is stopped if its potential is increased to its **stopping potential**, V_s. At this potential,

$$eV_s = hf - \phi$$

Therefore, a graph of V_s (on the y-axis) against f is a straight-line graph with:

- $\dfrac{h}{e}$ for its gradient m
- $f_0 = \dfrac{\phi}{h}$ for its x-intercept.

Summary questions

$e = 1.6 \times 10^{-19}$ C, $h = 6.63 \times 10^{-34}$ Js, $c = 3.00 \times 10^8$ ms^{-1}

1. Light of wavelength 460 nm is directed at a metal surface that has a work function of 2.1 eV. Calculate:
   a. the energy of a photon of wavelength 460 nm *(2 marks)*
   b. the maximum kinetic energy of the emitted photoelectrons. *(3 marks)*

> **Revision tip**
> Increasing the intensity of the incident radiation increases the number of photoelectrons per second, provided $f > f_0$.

2.4 Matter waves

De Broglie's hypothesis states that all matter particles have a wave-like nature and that the particle momentum mv is linked to its de Broglie wavelength λ by the equation

$$mv \times \lambda = h \text{ where } h \text{ is the Planck constant}$$

Combining the de Broglie equation with $\frac{1}{2}mv^2 = eV_A$ gives $\lambda = \dfrac{h}{\sqrt{(2meV_A)}}$.

The hypothesis was verified by the discovery that electrons in a beam are diffracted when they pass through a very thin metal foil.

> **Synoptic link**
> Electron diffraction here is referred to as 'low-energy' electron diffraction to distinguish it from the diffraction of electrons of much higher energies by nuclei. See Topic 26.9, Nuclear radius.

The transmission electron microscope (TEM)

Electrons in a beam are scattered when they pass through the sample. Electromagnetic coils acting as 'magnetic lenses' focus the scattered electrons onto a fluorescent screen to form a magnified image of the sample structure.

Diffraction at the lens apertures reduces the amount of detail in the image. Increasing the anode pd gives better resolution because it reduces the de Broglie wavelength of the electrons. The amount of detail is also affected by sample thickness and lens aberrations.

> **Revision tip**
> Make sure you know the position and function of three magnetic lenses (condenser, objective, and magnifier) in the TEM.

> **Synoptic link**
> See Topic 5.6, Diffraction.

The scanning tunnelling microscope (STM)

In the STM, electrons 'tunnel' across the small gap (≈ 1 nm) between a fine metal tip and the surface it scans.

- The tip is at a potential of about -1 V relative to the surface so that electrons only tunnel in one direction.
- Piezoelectric transducers are used to control the position of the tip.

In constant height mode, the tunnelling current varies as the tip scans across the surface and the gap width changes. The tunnelling current is recorded and used to map the height of the surface on a computer screen.

> **Revision tip**
> In constant current mode, the gap is kept constant by feedback to the piezoelectric transducer that controls the tip height.

Turning points in physics

Summary questions

$e = 1.6 \times 10^{-19}$ C, $h = 6.63 \times 10^{-34}$ Js, $m_e = 9.11 \times 10^{-31}$ kg, $m_p = 1.67 \times 10^{-27}$ kg

1. **a** Calculate the speed and de Broglie wavelength of an electron in an electron beam that has been accelerated from rest through a pd of 3500 V. *(4 marks)*
   b In a transmission electron microscope, state and explain how the image of a thin sample would change if the sample was moved so that the beam passed through a thinner part of the sample. *(3 marks)*

CHAPTER 3 SPECIAL RELATIVITY

3.1 About motion

Electromagnetic waves were thought to be vibrations in an invisible substance called **ether**. Detection of the ether was thought possible by comparing the time taken by light to travel the same distance in perpendicular directions. Michelson and Morley designed the interferometer in Figure 1 to measure such small differences.

The observer sees a pattern of interference fringes because of the difference in the path lengths of the two beams. Turning the apparatus through 90° in a horizontal plane swaps the beam directions relative to the Earth's motion through the ether. Ether theory predicted the difference in the travel times of the two beams would reverse, causing a measurable shift in the fringes. However, no fringe shift was observed. This 'null' result disproved the 'ether' theory.

▲ **Figure 1** *The Michelson-Morley interferometer*

3.2 Einstein's theory of special relativity

Einstein based his theory of special relativity on two postulates:

- **Physical laws have the same form in all inertial frames of reference** (frames of reference that move at constant velocity relative to each other).
- **The speed of light in free space, c, is invariant** (i.e., is always the same and is independent of the motion of the light source and the observer).

Relativistic effects

Time dilation: the time t between two events measured by a moving observer is stretched out ('dilated') compared with the proper time t_o (the time interval between the two events measured by an observer at rest relative to the events).	$t = \dfrac{t_o}{\sqrt{1 - \dfrac{v^2}{c^2}}}$
Length contraction: the length L of a rod moving in the same direction as its length is shorter than its proper length L_o (its length measured at rest).	$L = L_o \sqrt{1 - \dfrac{v^2}{c^2}}$
Relativistic mass: the mass m of a moving object is greater than its rest mass m_o and increases with speed. No material object can ever reach the speed of light as its mass would become infinite.	$m = \dfrac{m_o}{\sqrt{1 - \dfrac{v^2}{c^2}}}$
Mass and energy: Energy E and mass m are equivalent (i.e., interchangeable) on a scale given by $E = mc^2$	kinetic energy $E_k = mc^2 - m_o c^2$ $\dfrac{1}{2}mv^2$ at speeds $v \ll c$

Summary questions

1. **a** Explain the formation of a dark fringe in the fringe pattern observed when looking through the telescope in the Michelson-Morley experiment. *(3 marks)*
   b Describe what observation was expected when the apparatus was rotated through 90°. *(1 mark)*
   c State what they discovered and explain why their discovery was important. *(1 mark)*

Revision tip

As speed $v \to c$, m increases gradually to about $2m_o$ at $v \approx 0.9c$ and then increases sharply and $\to$ **infinity** as v approaches c.

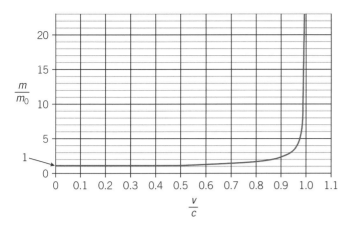

▲ **Figure 2** *Relativistic mass against speed*

Bertozzi's experiment

A linear accelerator was used to accelerate pulses of electrons from rest through a measured pd, V. He measured their speed v by timing each pulse between two electrodes a measured distance apart. A plot of $y =$ kinetic energy E_K against $x = \dfrac{v}{c}$ gives a curve that matches Figure 2 (whereas E_K against $\dfrac{1}{2}mv^2$ gives a much lower curve) that goes past $v/c = 1$.

Summary questions

$e = 1.6 \times 10^{-19}$ C, $c = 3.0 \times 10^8$ m s^{-1}

1. A beam of particles of rest energy 140 MeV travelling at a speed of $0.99c$ travels in a straight line between two stationary detectors which are 200 m apart. Calculate:

   a the proper time between a particle passing between two detectors. *(2 marks)*

   b the kinetic energy of each particle in MeV at this speed. *(3 marks)*

Electronics

▲ **Figure 1** *A MOSFET in use*

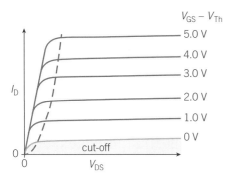

▲ **Figure 2** *Graph of I_D against V_{DS}*

Revision tip

The input resistance of a MOSFET (i.e., the resistance between the gate and the source) is very high. So the current between the gate and the source is negligible.

▲ **Figure 1** *The zener diode*

CHAPTER 1 DISCRETE SEMICONDUCTOR DEVICES
1.1 MOFSETs

An **n-channel enhancement MOSFET** (metal-oxide semiconducting field-effect transistor) is shown in the circuit in Figure 1. Figure 2 shows how the drain–source current, I_D, depends on the pd V_{DS} between the drain and the source for different values of the gate–source pd, V_{GS}.

- When $V_{GS} \le$ the **threshold voltage** V_{Th}, $I_D = 0$ and the transistor is OFF.
- When $V_{GS} > V_{Th}$, increasing V_{DS} from zero increases I_D linearly then the rate of increase decreases and I_D becomes constant (= saturated current I_{DSS}).

Summary questions

1. **a** Explain why the relay switch in Figure 1 closes when the input voltage is made sufficiently positive. *(3 marks)*
   b Sketch a graph to show how the drain–source current, I_D, for a MOSFET depends on the pd between the drain and the source, V_{DS}, when $V_{GS} > V_{Th}$. On your graph, mark the point P at which saturation occurs at minimum V_{DS}. *(3 marks)*

1.2 Zener diodes

Zener diode characteristics

A reverse-biased zener diode breaks down and conducts at a specific pd, its **breakdown voltage** V_Z. See Figure 1. A zener diode in series with a resistor R and a dc supply V_S is a **constant voltage source** as the output pd across the diode does not change when the current supplied to the output changes. The resistance of R is chosen between two limits:

- $R_{min} = \dfrac{(V_S - V_Z)}{I_{max}}$ where $I_{max} = \dfrac{\text{maximum power rating } P}{V_Z}$
- $R_{max} = \dfrac{(V_S - V_Z)}{I_{out}}$ where I_{min} = least current needed to operate the device at the output.

Summary questions

1. A 6.0 V, 300 mW zener diode is reverse-biased and in series with a 60 kΩ resistor and a 9.0 V battery.
   A 6.0 V, 30 mA lamp is connected across the zener diode. Calculate the current in:
   a the resistor *(2 marks)*
   b the zener diode. *(1 mark)*

1.3 Photodiodes
Photodiode characteristics

A photodiode conducts when it is reverse-biased and light is incident on it. Figure 1 shows how the current varies with the pd across the photodiode for different light intensities. Note that the dark current (i.e., at zero light intensity) is negligible.

Uses of a photodiode

- to monitor light intensity changes
- to convert light pulses into electronic pulses
- to detect atomic particles when used with a scintillator.

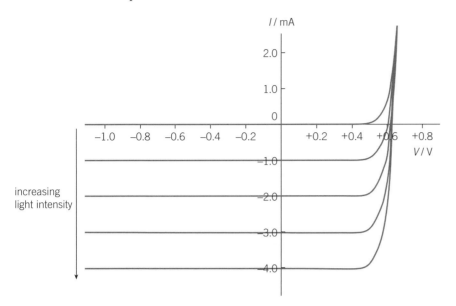

▲ **Figure 1** *A load line on the I against V graph for a photodiode*

1.4 Hall effect-sensors

The sensor is designed to detect, monitor, and measure magnetic fields. Charge carriers moving through the sample are forced by a magnetic field to one edge of the sample. The pd across the sample, the Hall voltage, V_H, is proportional to:

- the current I through the sample
- the component of the magnetic flux density B perpendicular to the sample.

Applications

- as a **proximity sensor**
- to monitor **'attitude'** or orientation of objects in a magnetic field relative to the field direction.

CHAPTER 2 SIGNALS AND SENSORS
2.1 Analogue and digital signals

Advantages of digital transmission

1. Elimination of noise by using regenerator amplifiers.
2. Digital signals can be compressed by using pulses of shorter duration so more bits per second can be sent.

Pulse code modulation is used to convert an analogue to a digital signal at a single output in a 3 stage process.

1. **Sampling** to produce a 'pulse amplitude modulated' (PAM) signal.
2. **Conversion to a data byte** (quantisation) on parallel outputs using an A/D converter.
3. **Parallel to serial conversion** to produce a sequence of bits at a single output.

Quality of conversion (back to an analogue signal)

1. **The frequency of the sampling process** (i.e., the sampling rate) must be at least twice the highest frequency of the analogue signal.

Summary questions

1. A reverse-biased photodiode is connected in series with a 9.0 V battery, a 4.7 kΩ resistor, and a milliammeter.
   a. When light at a certain intensity is incident on the photodiode, the milliammeter reads 1.2 mA. Calculate the pd across the photodiode. *(2 marks)*
   b. Calculate the photodiode pd when the light intensity is reduced by 50%. *(2 marks)*

Summary questions

1. a. State the physical property that is measured by a Hall effect sensor. *(1 mark)*
   b. When a Hall effect attitude device on the ground is orientated so its z-axis is vertical, it gives x, y, z readings of 3 µT, 4 µT, and 5 µT respectively.
      i. Calculate the magnitude of the magnetic flux density B. *(2 mark)*
      ii. Calculate the angle between the z-axis and the resultant magnetic field. *(1 mark)*

Electronics

Summary questions

1. An audio signal with a maximum amplitude of 6.4 V is converted into a digital signal. Each sample of the audio signal is converted into an 8-bit byte.
   a. What is the least change of the audio voltage that would produce a different byte? *(1 mark)*
   b. What byte is produced when the voltage of the audio signal is 4.1 V? *(2 marks)*

Revision tip
Remember that ntc thermistors have a resistance that falls with increase of temperature.

Synoptic link
See Topic 13.5, The potential divider, for more about thermistors and LDRs in potential dividers.

▲ **Figure 1** *(2.2.1)*

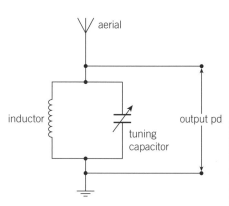

▲ **Figure 1** *A parallel LC circuit*

2. **The more bits in a sample** (i.e., the word-length), the smaller the spacing between adjacent levels and the greater the quality of the recovered analogue signal.

The number of bits transmitted per second = the sample frequency × the number of bits per sample.

2.2 Sensors
2.2.1 Sensor-operated systems

An electronic sensor contains a **sensing device** which supplies a signal to a **processing unit** which can operate an **output device** such as an indicator lamp or a relay.

Sensing devices in potential dividers

A potential divider that includes a thermistor or an LDR (light dependent resistor) can be used as a device to 'sense' temperature or light intensity changes.

Summary questions

1. An LDR and a 220 kΩ resistor R are connected in series with each other and a 12V power supply, as shown in Figure 1. The LDR has a dark resistance of 900 kΩ and a resistance of 50 kΩ in a darkroom when the room lights are on. Calculate the output pd from this circuit in:
   a. darkness *(2 marks)*
   b. in the dark room when the room lights are on. *(2 marks)*

CHAPTER 3 ANALOGUE SIGNAL PROCESSING
3.1 LC resonance filters
The parallel *LC* circuit

Figure 2 shows how the rms pd across the parallel LC circuit in Figure 1 varies with frequency.

The pd peak occurs at **the resonant frequency** $f_{RES} = \dfrac{1}{2\pi\sqrt{(LC)^{\frac{1}{2}}}}$.

When a radio or TV receiver is 'tuned in' to a channel, the resonant frequency of the *LC* circuit is adjusted to the channel's radio frequency by changing the inductance L or the capacitance C.

- **The bandwidth** f_B of the circuit is defined as the difference between the frequencies (either side of the peak) at $\sqrt{2}$ of the maximum pd (i.e., the 50% energy points of the curve).
- **The Q factor** $= \dfrac{f_{RES}}{f_B}$.

Summary questions

1. a. Calculate the capacitance of the capacitor in a parallel *LC* circuit designed to detect radio signals of frequency 1.5 MHz if it has a 0.25 mH inductor. *(2 marks)*
   b. Describe what is meant by the *Q* factor of a parallel *LC* circuit. *(2 marks)*

3.2 Ideal operation amplifiers

An operational amplifier has an inverting input, P, and a non-inverting input, Q. For pds V_- and V_+ applied to inputs P and Q respectively,

the output pd $\quad V_{OUT} = A_0(V_+ - V_-)$ **(the transfer function)**

where A_0 is the **open-loop gain** (typically of the order of 10^5).

1. The output pd cannot exceed the limits determined by the power supply. For a ±15 V supply, $(V_+ - V_-) < \pm 150\,\mu V$ if the output pd is not to be saturated (i.e., not reach ±15 V).
2. If either P or Q are earthed, the terminal that is not earthed is said to be at **virtual earth**, provided the output pd is not saturated.
3. An ideal op-amp has an infinite input resistance and zero output resistance.

The comparator

If the output pd V_{OUT} is:

- at positive saturation, then $V_+ > V_-$, (i.e., pd at Q > pd at P)
- at negative saturation, then $V_- > V_+$ (i.e., pd at P > pd at Q).

▲ **Figure 2** *The frequency response of a parallel LC circuit*

Revision tip
Make sure you can explain how the comparator is used to switch a device (e.g., a buzzer or an LED on or off) in response to a change of pd from a sensor in a potential divider.

Summary questions

1. **a** Explain, in relation to an operational amplifier, what is meant by virtual earth. *(2 marks)*
   b **i** When the temperature of the thermistor in Figure 1 decreases, explain why the buzzer switches on at a certain temperature. *(3 marks)*
   ii Describe the adjustment you must make to the variable resistor so the buzzer switches on at a higher temperature. *(1 mark)*

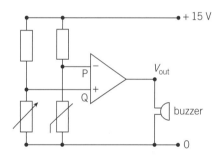

▲ **Figure 1** *A temperature-operated buzzer*

3.3 Operational amplifiers circuits

Feedback is the process of taking some of the output signal from an amplifier and adding it to the input signal. Positive feedback increases the output and negative feedback reduces it. Resistors are used to supply negative feedback to reduce the gain of an op-amp.

The inverting amplifier

$$\text{Voltage gain } \frac{V_{out}}{V_{in}} = -\frac{R_F}{R_{in}}$$

where R_F is the feedback resistor and R_{in} is the input resistor.

The summing amplifier

The circuit is like Figure 1 but with three input terminals and three input resistors.

$$\text{The output pd } V_{out} = -IR_F = -\left(\frac{V_1}{R_1} + \frac{V_2}{R_2} + \frac{V_3}{R_3}\right)R_F$$

The non-inverting amplifier

$$\text{Voltage gain } \frac{V_{out}}{V_{in}} = 1 + \frac{R_F}{R_{in}}$$

where R_F is the feedback resistor and R_{in} is the input resistor.

The frequency response of an operational amplifier

For alternating input voltages, the voltage gain is constant up to a certain frequency and decreases at higher frequencies.

▲ **Figure 1** *The inverting amplifier*

Revision tip
The voltage gain of the inverting amplifier is independent of the open-loop voltage gain, A_0.

Electronics

▲ Figure 2 *The non-inverting amplifier*

Revision tip
You need to know how to derive the voltage gain equation for the inverting amplifier only. See the notes on the Electronics option in the Kerboodle resources.

The bandwidth is the frequency range over which the voltage gain is constant.

The bandwidth × the voltage gain = constant.

Summary questions

1 **a** A sinusoidal alternating voltage with a peak voltage of 5.0 V is applied to the input of a non-inverting ±15 V operational amplifier which has a voltage gain of ×4. Draw graphs on the same axes to show how the input pd and the output pd vary with time. *(3 marks)*

   b A summing amplifier has a 100 kΩ feedback resistor and four input terminals A to D with input resistances of 100 kΩ, 200 kΩ, 400 kΩ, and 800 kΩ respectively. A 4-bit byte 1011 is applied to inputs DCBA respectively where 1 represents a 4 V input. Calculate the output voltage. *(2 marks)*

Summary questions

1 **a** Draw a truth table for the Boolean expression $C = \overline{\overline{A}.\overline{B}}$ *(4 marks)*

   b Hence show that $\overline{\overline{A}.\overline{B}} = A + B$ *(1 mark)*

CHAPTER 4 DIGITAL SIGNAL PROCESSING

4.1 Logic gates

Make sure you know the symbols and truth tables of NOT, OR, AND, NOR, EOR, and NAND gates. **In Boolean algebra**, + represents OR and a dot represents AND. Also, a bar above an input or output symbol represents NOT. For example, Q = (A + B).C means output Q = 1 if input A OR input B = 1 AND input C = 1.

4.2 Sequential logic

Binary counter See Figure 1 and Table 1	A series of identical circuits called T-type flip flops	Counts up repeatedly from 0000 to 1111 as the falling edge (1→0) at each Q output causes the next Q to switch.	A binary counter with n flip-flops goes through 2^n output states (the counter's modulus) before it resets itself.
BCD counter	Resets to 0 when the binary count reaches 1010 (=decimal 10)	The inputs of a 2-input AND gate to Q_4 and Q_2 and the AND output to the reset terminal.	Can be modified to count up to any number <10 by connecting the appropriate Q outputs to a 2 or 3 input AND gate which has its output gate are connected to the reset pin.
Johnson counter See Figure 2 and Table 2	A series of D-type flip flops with the last $\overline{Q}$ output connected to the first D input.	Each Q output is shifted to the next output when a clock pulse is applied	The clock pulses are applied, after the Q outputs have been reset to 0. The sequence of changes shown in Table 2 takes place repeatedly.

(a) circuit (reset terminals not shown)

(b) input and output states

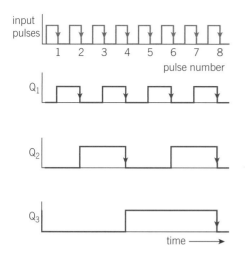

pulse number	Q_3	Q_2	Q_1
0	0	0	0
1	0	0	1
2	0	1	0
3	0	1	1
4	1	0	0
5	1	0	1
6	1	1	0
7	1	1	1
8	0	0	0

▲ **Figure 1** *A 3-bit binary counter (reset terminals not shown)*

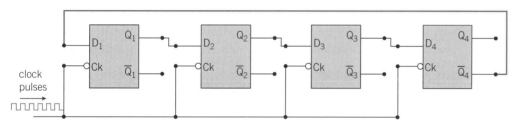

▲ **Figure 2** *A Johnson counter (reset terminals not shown)*

The astable oscillator

Clock pulses are provided by an **astable** oscillator. This is a digital circuit that repeatedly switches its output voltage between two voltage levels, as shown in Figure 3.

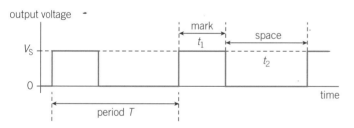

▲ **Figure 3** *The astable multivibrator output voltage*

The **mark-to-space ratio** (MSR) of the waveform $= \dfrac{t_1}{t_2}$

The **duty cycle** $= \dfrac{t_1}{T} \times 100\% \left(= \dfrac{\text{MSR}}{(\text{MSR} + 1)} \times 100 \right)$

The frequency and the mark-to-space ratio are determined by the capacitance and resistance values of the capacitor and resistors in the oscillator.

CHAPTER 5 DATA COMMUNICATION SYSTEMS
5.1 Systems and links

The **bandwidth** of a transmission path is the range of frequencies that can be transmitted. The bandwidth is divided into frequency **channels** with a different signal allocated to each channel. For digital signals, each channel can carry a number of digital signals using **time-division multiplexing** (TDM). In this process, the duration of each bit is reduced and the time between one bit of a signal and the next is divided into a number of equal time slots which are used to carry bits from other signals.

▼ **Table 2** *A Johnson counter sequence*

Input pulse number	Q_1	Q_2	Q_3	Q_4
0	0	0	0	0
1	1	0	0	0
2	1	1	0	0
3	1	1	1	0
4	1	1	1	1
5	0	1	1	1
6	0	0	1	1
7	0	0	0	1
8	0	0	0	0

Summary questions

1. **a** Explain how a binary counter can be adapted so it counts from 0 to 5 repeatedly. *(3 marks)*
   b Use Figure 2 to explain how a Johnson counter with 3 flip-flops works. *(4 marks)*

Synoptic link

See Topic 5.3, Total internal reflection for more about optical fibres.

Electronics

Summary of the advantages and disadvantages of each type of communication link

Type of link	Twisted wire pair	TV coaxial cable	Radio/microwave link	Optic fibre
Typical attenuation	≈5× weaker at 10 MHz per 100 m	≈10× weaker at 100 MHz per 100 m	≈10^8× weaker at 1 GHz per km	≈0.95× weaker for infrared per km
Bandwidth	500 kHz	50 MHz	FM 200 kHz, Microwaves 1 GHz	≈GHz for 1 km (bandwidth decreases with increasing distance)
Noise	electrical interference, cross-over from other wires	electrical interference from electric motors	electrical storms, cosmic sources	no electrical interference, noise removed by regenerators
Bit rate /10^6 bits per second	10	100	Digital TV 20, Microwave beam 70	>10^6
Repeater/regenerator spacing	short distances	up to 2–3 km	line of sight for microwaves	up to 100 km or more
Security	low, unless coded digital signals used			high

Note: Radio waves of frequencies < 30 MHz reflect from the ionosphere in the upper atmosphere back to the ground (and back to the ionosphere etc.).

5.1 Summary questions

1. **a** Which type of communication link is suitable for:
   i ship to shore communications *(1 mark)*
   ii signals to and from a satellite? *(1 mark)*
   b A digital signal consists of 16-bit words transmitted every 0.80 ms. Each bit is a pulse lasting 0.50 μs. How many such signals could be transmitted using time-division multiplexing in a single frequency channel and what would be the bit rate in the channel? *(3 marks)*

5.2 Modulation

Analogue signals are transmitted by **amplitude modulation (AM) or frequency modulation (FM)** where the carrier amplitude or frequency is modulated by the signal amplitude.

For a carrier wave of frequency f_c modulated by waves of frequency f_m, the **bandwidth** of the signal:

- = $2 f_m$ **for amplitude modulation**
- = $2 (\Delta f + f_M)$ **for frequency modulation** (where Δf is the maximum deviation of the sidebands from the carrier frequency).

The relative advantages of AM and FM transmission

Quality: FM less affected by noise than AM

Range: AM range greater than FM range for the same power

Relative advantages of Pulse Code Modulation (PCM)

- no noise and higher quality
- more than one signal can be transmitted using TDM on a single frequency channel
- a PCM signal can be security-coded.

The main disadvantages of PCM is its greater bandwidth.

5.2 Summary questions

1. **a** A radio transmitter broadcasts a frequency-modulated signal of bandwidth 500 kHz at a carrier frequency of 98.0 MHz. Explain what is meant by the bandwidth of a signal and calculate the upper and lower limits of the allocated frequency channel. Assume only the lower frequency band is used. *(2 marks)*
   b An FM radio station transmits its signal in a frequency channel of bandwidth 200 kHz at a carrier frequency of 102.25 MHz. The upper limit of the modulation frequency is 15 kHz. Calculate:
   i the maximum deviation of the signal frequency from the carrier frequency *(1 mark)*
   ii the carrier frequency of the next higher frequency channel. *(1 mark)*

Answers to practice questions

Chapter 17

1. **A**: Speed of J = circumference/time for 1 orbit = $2\pi R_J/T_J$

   Speed of E = circumference/time for 1 orbit = $2\pi R_E/T_E$

   $\dfrac{\text{Speed of J}}{\text{Speed of E}} = \dfrac{2\pi R_J/T_J}{2\pi R_E/T_E} = \dfrac{R_J}{R_E} \times \dfrac{T_E}{T_J} \approx 5 \times \dfrac{1}{12} = 0.42$ [1]

2. **C**: $E_K = \tfrac{1}{2}mv^2 = mr$
   $a\quad v^2/r\quad 2$ [1]

3. **A**: Support force $S = mg - ma = mg - \dfrac{mv^2}{2}$ (as $mg - S = ma$) [1]

4. **a i** Frequency of rotation = 1400/60 s = 23.3 Hz
   Angular speed $\omega = 2\pi f = 2\pi \times 23.3$ Hz = 147 rad s^{-1} = 150 rad s^{-1} (2 s.f.) [1]

   ii speed of point on wall = ωr = 147 rad s^{-1} × 0.45 m = 66 m s^{-1} [1]

   iii centripetal acceleration = $\omega^2 r$ = (147 rad s^{-1})^2 × (½ × 0.45 m) = 4900 m s^{-2} [1]

   b The clothing in the tub cannot leave the tub as it is forced to move round in circular motion by the wall and it cannot pass through the holes in the wall [1]

   The water in the clothing passes through the holes in the wall because the clothing cannot provide enough force on the water to retain it. [1]

5. **a** The speed of a point on the tyre $v = \omega R = 2\pi f R$ where $R = 0.60$ m ÷ 2 = 0.30 m

   So the frequency of rotation $f = v/2\pi r$ = 25 m s^{-1}/(2π × 0.30 m) = 13.3 Hz (2 s.f.) [1]

   b i Centripetal force $F = m\omega^2 r$ where $r = 0.21$ m and $\omega = 2\pi f = 2\pi \times (26.5$ Hz$) = 167$ rad s^{-1} [1]

   Therefore $F = 0.025$ kg × (167 rad s^{-1})^2 × 0.21 m = 146 N = 150 N (2 s.f.) [1]

   ii Graph; (parabolic) curve through the origin with increasing gradient [1]

   passes through $F = 150$ N, $v = 25$ m s^{-1} [1] and through a further point consistent with $F \propto V^2$ [1]

6. **a** The speed of a point on the tyre $v = \omega r = 2\pi f r$ where $r = 0.50$ m ÷ 2 = 0.25 m

   So the frequency of rotation $f = v/2\pi r$ = 8.2 m s^{-1}/(2π × 0.25 m) = 5.2 Hz [1]

   b Centripetal acceleration = v^2/r = (8.2 m s^{-1})^2/0.25 m = 270 m s^{-2} [1]

7. **a** Use a stopwatch to time 20 complete turns of the turntable (by observing an object fixed to the turntable near its rim) [1]

   Repeat the procedure 3 times and obtain the average timing; the time period is the average timing/20. [1]

   b i Angular displacement $\Delta\theta$ in 180 s = angular speed × time = 0.86 rad s^{-1} × 180 s = 155 rad. [1]
   Number of turns = $\Delta\theta/2\pi$ = 24.7 turns = 25 turns (2 s.f.) [1]

   ii Centripetal acceleration = $\omega^2 r$ = (0.86 rad s^{-1})^2 × (0.270 m/2) = 0.10 m s^{-2} [1]

8. **a i** Loss of PE = mgh = 2200 kg × 9.81 m s^{-2} × 61 m = 1.3(2) × 10^6 J [1]

   ii Assuming gain of KE = loss of PE then
   $\tfrac{1}{2}mv_{max}^2 = 1.32 \times 10^6$ J [1]
   Therefore $v_{max}^2 = (2 \times 1.32 \times 10^6$ J$)/2200$ kg = 1200 m^2 s^{-2}
   hence $v_{max} = 34.6$ m s^{-1} = 35 m s^{-1} (2 s.f.) [1]

   b Centripetal force mv^2/r = (2200 × (34.6 m s^{-1})^2/52 m = 50 600 N [1]
   Support force due to the track $S = mg + mv^2/r$ = (2200 × 9.81 m s^{-2}) + 50 600 N = 72 000 N [1]

Chapter 18

1. **D**: The maximum speed = $A\omega$ and $\omega = \sqrt{(k/m)}$ [1]

2. **C**: A is incorrect because T does not change as the amplitude decreases
   B is incorrect because any damped system gradually returns to equilibrium
   D is incorrect because any system that is lightly damped can be made to resonate [1]

3. **B**: A is incorrect because E_K decreases
   C is incorrect because E_P increases
   D is incorrect because E_K increases and E^P decreases [1]

4. **a** $T = 35$ s/20 = 1.75 s so $\omega = 2\pi/T = 2\pi/1.75$ s = 3.59 rad s^{-1} [1]
   For $A = 22$ mm, acceleration $a = -\omega^2 A$
   = $-(3.59$ rad s$^{-1})^2 \times 0.022$ m = -0.28 m s^{-2} [1]
   magnitude of maximum acceleration = 0.28 m s^{-2} [1]

5. **a i** 40 mm [1]

   ii Angular frequency $\omega = 5$ rad s^{-1} therefore $T = 2\pi/\omega = 2\pi/5 = 1.26$ s [1]

   b At $t = 0.90$ s, displacement $x = 40$ mm × cos $5t$ = 40 mm × cos (5 × 0.90 s) = -8.4 mm [1]

6. Extension $e = (0.320$ m $- 0.250$ m$) = 0.070$ m
   The spring constant $k = mg/e$ = (0.15 kg × 9.81 m s^{-2})/0.070 m = 21 N m^{-1} [1]

   Therefore $T = 2\pi\sqrt{\dfrac{0.15\text{ kg}}{21\text{ N m}^{-1}}} = 0.53$ s [1]

   Alternative method $mg = ke$ therefore $m/k = e/g$ [1]

   hence $T = 2\pi\sqrt{\dfrac{e}{g}}$ [1]

   $= 2\pi\sqrt{\dfrac{0.070\text{ m}}{9.81\text{ m s}^{-2}}} = 0.53$ s [1]

Answers to practice questions

7 a See Topic 18.5, Figure 2 Light damping curve: the time period should be unchanged as the amplitude decreases [1]
the amplitude should decrease exponentially [1]

b If the amplitude decreases by a factor f in 1 cycle, it decreases by a factor f^5 in 5 cycles therefore
$f^5 = 0.45$ [1]
which gives $f = 0.85$ by trial and error
(or by calculation ;– $\ln f = (\ln 0.45)/5 = -0.160$
which gives $f = e^{-0.160} = 0.85$ [1])
Total energy $\propto A^2$ so total energy decreases by a factor $f^2 = 0.72$ or 72% [1]

8 a The periodic force due to the bumps has a time period of $4.0\,\text{s}$ (= $12\,\text{m}/3.0\,\text{m s}^{-1}$) and therefore the time period of the periodic force is $4.0\,\text{s}$ [1]

The carrier box oscillates at resonance due to the periodic force so the time period of the oscillations at the natural frequency of the trailer is $4.0\,\text{s}$ [1]

As there are two identical springs in parallel, the spring constant of the system = $2k$ where k is spring constant of each spring. Therefore the time period $T = 2\pi\sqrt{m/2k}$ [1]
Rearranging $T = 2\pi\sqrt{m/2k}$ gives $k = 2\pi^2 m/T^2$ where $T = 4.0\,\text{s}$ [1]
Therefore $k = 2\pi^2 \times 210\,\text{kg}/(4.0\,\text{s})^2$ [1] = $260\,\text{N m}^{-1}$ [1]

b No, the reduction in mass means the time period of natural oscillations will be less [1]
so the trailer will only resonate over these speed bumps at a higher speed [1]
as the time between successive bumps would need to be less. [1]

Chapter 19

1 B: Work is done on the air so D is incorrect. There is no change in the internal energy as the temperature at the end is the same as at the start so B must be correct. [1]

2 B: Energy is needed to melt a substance so A is incorrect. B is correct because the specific heat capacity is inversely proportional to the rate of increase of temperature and this is higher for the solid than the liquid (which means C is incorrect). The internal energy increases during melting so D is incorrect. [1]

3 D: Specific latent heat unit = J kg^{-1} = N m kg^{-1} = $(\text{kg m s}^{-2})\,\text{m kg}^{-1}$ = m^2s^{-2} [1]

4 a Energy to heat mass m by temperature rise $\Delta T = mc\Delta T$
Energy supplied = IVt
Assuming no energy loss to the surroundings, $mc\Delta T = IVt$ [1]
Therefore specific heat capacity $c = IVt/m\,\Delta T$ = $2.60\,\text{A} \times 12.2\,\text{V} \times 200\,\text{s}/(1.210 \times (33.5 - 18.0)\,°\text{C})$ [1] = $338\,\text{J kg}^{-1}\text{K}^{-1}$ [1]

b i % uncertainty in current = $(0.05/2.60) \times 100\% = 1.9\%$
% uncertainty in pd = $(0.10/12.2) \times 100\% = 0.8\%$
% uncertainty in time = $(0.40/200) \times 100\% = 0.2\%$ [1 for all above correct]
Therefore % uncertainty in energy supplied = $(1.9 + 0.8 + 0.2)\% = 2.9\%$ [1]

ii % uncertainty in temperature change = $(1.0/15.5) \times 100\% = 6.5\%$ [1]

iii % uncertainty in mass = $(0.005/1.210) \times 100\% = 0.4\%$
% uncertainty in c = sum of above % measurements uncertainties = 9.8% [1]

c Measure the temperature at intervals after the heater is switched off [1]. If there is no energy loss, the temperature should be unchanged after the heater is switched off [1].

5 Let m = mass of ice melted, m_1 = mass of water in the calorimeter, m_2 = mass of the copper can, the specific heat capacity of water = c_1, and the specific heat capacity of copper = c_2
Energy needed to melt the ice = ml = $3.40 \times 10^5\,\text{J kg}^{-1} \times m$ [1]
Energy needed to raise the temperature of the melted ice to 12°C = $mc_1(T - 0) = (4200\,\text{J kg}^{-1}\text{K}^{-1} \times 12\,\text{K} \times m) = 5.04 \times 10^4\,\text{J kg}^{-1} \times m$ [1]
Total energy needed
= $(3.40 \times 10^5 + 5.04 \times 10^4)\,\text{J kg}^{-1}\,m$
= $3.90 \times 10^5\,\text{J kg}^{-1}\,m$
Energy released by the water and can = $m_1c_1(T_o - T) + m_2c_2(T_o - T)$ [1] = $[(0.080\,\text{kg} \times 390\,\text{J kg}^{-1}\text{K}^{-1} \times (15 - 12)\,\text{K}] + [(0.120\,\text{kg} \times 4200\,\text{J kg}^{-1}\text{K}^{-1} \times (15 - 12)\,\text{K}]$ = 94 J + 1512 J = 1606 J [1]
Assuming energy needed = energy released, $5.04 \times 10^4\,\text{J kg}^{-1} \times m = 1606\,\text{J}$ [1]
Therefore $m = 1606\,\text{J}/3.90 \times 10^5\,\text{J kg}^{-1} = 4.1 \times 10^{-3}\,\text{kg}$ [1]

6 a Energy transferred per second by the water = $(m/t)c(T_2 - T_1)$ = $0.042\,\text{kg s}^{-1} \times 4200\,\text{J kg}^{-1}\text{K}^{-1} \times (50°\text{C} - 44°\text{C})$ [1] = $1060\,\text{J s}^{-1}$ = $1100\,\text{J s}^{-1}$ (2 s.f.) [1]

b The average temperature of the radiator when it is hot = 47°C [1]
Energy E needed to heat the radiator from 15°C to 47°C = $m_{rad}\,c_{rad}\,\Delta T$ = $28\,\text{kg} \times 560\,\text{J kg}^{-1}\text{K}^{-1} \times (47°\text{C} - 15°\text{C}) = 5.0 \times 10^5\,\text{J}$ [1]
In time t, energy transferred $E = Pt$ where P is the mean rate of energy transfer during the time the radiator heats up. Assume 50% of the energy supplied in this time is transferred to the surrounding and 50% is used to heat the radiator then $P = \frac{1}{2} \times 1060\,\text{J s}^{-1} = 530\,\text{J s}^{-1}$. [1]
Rearranging $E = Pt$ gives $t = E/P$ = $5.0 \times 10^5\,\text{J}/530\,\text{J s}^{-1} = 940\,\text{s}$. [1]

7 a The heater supplied energy at a constant rate. If the solid was fully-insulated (from the surroundings), its temperature would have increased at a constant rate. [1]

Because it was not fully insulated, energy was transferred from the solid to the surroundings by heating. The rate of transfer of energy would therefore have been less than if it had been fully insulated. [1]

As the solid became hotter, the rate of transfer of energy to the surroundings increased so its rate of increase of temperature decreases. [1]

    b Initially the solid is at the same temperature as the surroundings so the rate of transfer to the surroundings is zero initially. [1]

Therefore, the energy per second supplied by the heater $P = m c (\Delta T/\Delta t)_o$ where $(\Delta T/\Delta t)_o$ = its initial rate of increase of temperature. [1]

From the graph, $(\Delta T/\Delta t)_o \approx 6\,K/60\,s = 0.10\,K\,s^{-1}$ [1]

Therefore, the specific heat capacity of the solid $\approx P/m(\Delta T/\Delta t)_o = 50\,W/(0.58\,kg \times 0.10\,K\,s^{-1}) = 860\,kg\,s^{-1}$. [1]

    c The temperature would have increased for 300 s in the same way then it would have decreased [1] at a decreasing rate [1]

8 a Energy transferred to cool the water = $m c \Delta T$ = $0.070\,kg \times 4200\,J\,kg\,K^{-1} \times 10\,K = 2940\,J$ [1]

Rate of transfer of energy = $2940\,J/140\,s = 21\,J\,s^{-1}$ [1]

  b i Energy transferred to freeze the water = $m L$ = $0.070\,kg \times 340\,kJ\,kg^{-1} = 23\,800\,J$ [1]

Time taken to transfer energy = $23\,800\,J/21\,J\,s^{-1} = 1130\,s = 1100\,s$ (2 s.f.) [1]

    ii The rate of energy transfer is assumed to be the same when it freezes as when it cools [1]. This may not be correct because the rate of transfer during cooling changes. [1]

When ice forms in the beaker and water is still present, the rate of transfer of energy from the beaker is unchanged, transferring energy at the same rate as water [1] Ice does/may not transfer energy at the same rate as water does [1]

Chapter 20

1 D: Rearranging $pV = nRT$ gives $n_p = pV/RT$ for the gas in P.

For the gas in Q, $n_q = 0.60\,pV/0.75\,TR = 0.80\,pV/RT = 0.80\,n$ [1]

2 B: Mass of gas m = number of moles n × molar mass M

Volume $V = nRT/p$ (from $pV = nRT$)

Density $= \dfrac{m}{V} = \dfrac{nM}{(nRT/p)} = \dfrac{pM}{RT}$ [1]

3 A: T = constant so molar mass m × mean square speed = constant $\propto T$

$\dfrac{\text{rms speed for X}}{\text{rms speed for Y}} = (800\,m\,s^{-1}/1600\,m\,s^{-1}) = 0.5$

$M_Y/M_X = \dfrac{\text{mean square speed for X}}{\text{mean square speed for Y}} = 0.5^2 = 0.25$

$M_Y = 0.25\,M_X = 0.25\,M$ [1]

4 a Graph should be a straight line with a +gradient through origin [2]

  b i as **a** with a steeper gradient [1]

    ii as **a** with a lesser gradient [1]

5 a i Rearrange $pV = nRT$ to give $n = pV/RT = 1.40 \times 10^5\,Pa \times 0.020\,m^3/(8.31\,J\,mol^{-1}\,K^{-1} \times 285\,K) = 1.18\,mol = 1.2\,mol$ (2 s.f) [2]

    ii Total KE $= \dfrac{3}{2}nRT = 1.5 \times 1.18 \times 8.31\,J\,mol^{-1}\,K^{-1} \times 285\,K = 4200\,J$ [1]

   iii Average KE of a molecule = $3/2\,kT = 1.5 \times 1.38 \times 10^{-23}\,J\,K^{-1} \times 285\,K = 5.9 \times 10^{-21}\,J$ [1]

5 b Internal energy U = total KE of molecules [1] so $\Delta U = 3/2\,nR\Delta T = 1.5 \times 1.18 \times 8.31\,J\,mol^{-1}\,K^{-1} \times (320 - 285)\,K = 510\,J$ [1]

6 a The rms speed of the gas molecules = the square root of the mean value of the square of the molecule speeds [1]

  b i See SB p53 Figure 1; curve from 0 to max number then decreasing [1] and tending to zero [1]

    ii See SB p53 Figure 2; same general shape as above but with a lower max [1] further along speed axis [1]

  c Mass of a nitrogen molecule = $0.028\,kg\,mol^{-1}/6.02 \times 10^{23}\,mol^{-1} = 4.65 \times 10^{-26}\,kg$ [1]

Mean kinetic energy $\dfrac{1}{2}mc_{rms}^2 = \dfrac{3}{2}kT = 1.5 \times 1.38 \times 10^{-23}\,J\,K^{-1} \times 273\,K = 5.65 \times 10^{-21}\,J$ [1]

Therefore rms speed = $(2 \times \text{mean KE}/m)^{1/2}$ = $(2 \times 5.65 \times 10^{-21}\,J/4.65 \times 10^{-26}\,kg)^{1/2} = 490\,m\,s^{-1}$ [1]

7 a i The molecules move about unpredictably, colliding with each other unpredictably and changing directions when they collide [1]

    ii An elastic collision is one in which the total kinetic energy after the collision is the same as before the collision [1]

  b i Initial momentum = $5.3 \times 10^{-26}\,kg \times 1200\,m\,s^{-1} = 6.36 \times 10^{-23}\,kg\,m\,s^{-1}$ [1]

Final momentum = $-6.36 \times 10^{-23}\,kg\,m\,s^{-1}$ [1]

Change of momentum $\Delta p = -6.36 \times 10^{-23}\,kg\,m\,s^{-1} - 6.36 \times 10^{-23}\,kg\,m\,s^{-1} = (-)1.27 \times 10^{-23}\,kg\,m\,s^{-1} = (-)1.3 \times 10^{-23}\,kg\,m\,s^{-1}$ (2 s.f.) [1]

    ii Force F = pressure × area [1] = $120 \times 10^3\,Pa \times 1 \times 10^{-6}\,m^2 = 0.12\,N$ [1]

F = no. of impacts per sec × change of momentum [1]

No. of impacts per second = $F/\Delta p = 0.12\,N/1.27 \times 10^{-23}\,kg\,m\,s^{-1} = 9.4 \times 10^{21}\,s^{-1}$ [1]

Answers to practice questions

Chapter 21

1 **C**: The distance r from the centre of the planet to a point at height R above the surface is $2R$. g is inversely proportional to r^2 so at $r = 2R$, $g = ¼ \times g_s$ so A and B are incorrect. V is inversely proportional to r so at $r = 2R$, $V = ½ \times V_s = ½ g_s R$ so D is incorrect. [1]

2 **A**: The unit of gravitational potential gradient is the same as the unit of gravitational field strength which is the same as the unit of acceleration. So the base unit combination for gravitational potential gradient is m s^{-2} [1]

3 **B**: V is inversely proportional to r so at $r = 2R$, $V = ½ \times V_s$ so $V/V_s = ½$. Curve B is therefore the correct curve as it passes through $V/V_s = ½$ at $r = 2R$. [1]

4 **a** The mass of the planet $m = \frac{4}{3}\pi R^3 \times$ density ρ

$g_s = \frac{GM}{R^2} = \frac{G \times \frac{4}{3}\pi R^3 \times \rho}{R^2} = \frac{4}{3}\pi G\rho R$ [1]

Rearranging this equation gives $\rho = \frac{g_s}{\frac{4}{3}\pi GR}$

$= \frac{1.62}{\frac{4}{3}\pi \times 6.67 \times 10^{-11} \times 1.74 \times 10^6}$ [1] $= 3330\,\text{kg m}^{-3}$ [1]

b $½ m v^2 \geq \frac{GMm}{r}$ gives escape speed $v_{esc} = \sqrt{\frac{2GM}{R}}$

$= \sqrt{\frac{2g_s R^2}{R}} = \sqrt{2g_s R}$ [1]

$= \sqrt{2 \times 1.62 \times (1.74 \times 10^6)} = 2370\,\text{m s}^{-1}$ [1]

5 **a** Rearranging $\frac{r^3}{T^2} = \frac{GM}{4\pi^2}$ gives $m = \frac{4\pi^2 r^3}{GT^2}$ where $T = 16.0 \times 24 \times 60 \times 60\,\text{s} = 1.38 \times 10^6\,\text{s}$

$M = \frac{4\pi^2 r^3}{GT^2} = \frac{4\pi^2 \times (1.22 \times 10^9)^3}{6.67 \times 10^{-11} \times (1.38 \times 10^6)^2}$ [1]

$= 5.64 \times 10^{26}\,\text{kg}$ [1]

b Volume $V = \frac{4}{3}\pi R^3 = \frac{4}{3}\pi (6.03 \times 10^7)^3 = 9.18 \times 10^{23}\,\text{m}^3$ [1]

density $= \frac{M}{V} = \frac{5.64 \times 10^{26}\,\text{kg}}{9.18 \times 10^{23}\,\text{m}^3} = 615\,\text{kg m}^{-3}$ [1]

c Let P be the point along the line between their centres at which the gravitational field strengths are equal and opposite.

For Saturn, $g_s = -\frac{GM_s}{r_1^2}$ where r_1 is the distance from the centre of Saturn to point P

For Triton, $g_T = -\frac{GM_T}{r_2^2}$ where r_2 is the distance from the centre of Triton to point P

At P, $g_s = g_T$ therefore $-\frac{GM_s}{r_1^2} = -\frac{GM_T}{r_2^2}$ [1]

Rearranging this equation gives $\frac{r_1}{r_2} = \sqrt{\frac{M_s}{M_T}} = \sqrt{\frac{5.64 \times 10^{26}\,\text{kg}}{1.35 \times 10^{23}\,\text{kg}}} = 64.4$ therefore $r_1 = 64.6 r_2$ [1]

Since $r_1 + r_2 = D$ where D = the distance between their centres = 1.22×10^9 m

Therefore $64.6 r_2 + r_2 = D$ so $65.6 r_2 = D$ [1]

$r_2 = D/65.6 = 1.22 \times 10^9\,\text{m}/65.6 = 1.86 \times 10^7\,\text{m}$ [1]

6 **a i** $F_{grav} = Gm^2/D^2 = mv^2/r$ [1] where r = radius of orbit = $D/2$ and v = orbital speed

Therefore $v^2 = GM/2D$ hence $v = \sqrt{\frac{GM}{2D}}$ [1]

ii Time period T = circumference/speed = $\pi D/v$

[1] $= \pi D \div \sqrt{\frac{GM}{2D}} = \pi \sqrt{(2D^3/GM)}$ [1]

b Squaring and rearranging $T = \pi \sqrt{(2D^3/GM)}$

gives $D^3 = \frac{GM}{2\pi^2} \times (80 \times 365 \times 24 \times 3600)^2$ [1]

$= 4.3 \times 10^{37}\,\text{m}^3$ [1]

Therefore $D = 3.5 \times 10^{12}\,\text{m} = 23$ AU [1]

7 For an extension of 180 mm, the spring is at or near its elastic limit and its tension would be about 0.9 N ($= 5.0\,\text{N m}^{-1} \times 0.180\,\text{m}$). [1] Using a mass of weight 0.9 N would give a difference in extension of 180 mm $\times 0.05\,\text{N kg}^{-1}/9.8\,\text{N kg}^{-1} = 0.92$ mm [1].

The extension is too small to measure accurately using the metre ruler. [1]

Using a significantly larger mass would stretch the spring beyond its limit of proportionality and the calculation would not be reliable as the spring constant would change with extension beyond the limit of proportionality. [1]

8 **a** Closest distance between Jupiter and Mars = $7.8 \times 10^{11}\,\text{m} - 2.3 \times 10^{11}\,\text{m} = 5.5 \times 10^{11}\,\text{m}$ [1]

$g_j = \frac{GM}{r^2} = -\frac{6.67 \times 10^{-11} \times 1.90 \times 10^{27}}{(5.5 \times 10^{11})^2}$

$= 4.2 \times 10^{-7}\,\text{N kg}^{-1}$ [1]

b $g_s = \frac{GM}{r^2} = -\frac{6.67 \times 10^{-11} \times 2.0 \times 10^{30}}{(2.3 \times 10^{11})^2}$

$= 2.5 \times 10^{-3}\,\text{N kg}^{-1}$ [1]

Resultant gravitational field strength = $g_s - g_j = 2.5 \times 10^{-3}\,\text{N kg}^{-1}$ towards the Sun [1]

Chapter 22

1 **B**: doubling the charge on T doubles the force and doubling the separation makes the force four times smaller. So the effect of both changes is to half the force (i.e., $F \to 2 \times ¼\,F = ½\,F$) [1]

2 **A**: a small + 'test' charge at P experiences a force F due to X away from X, a force F due to Z away from Z and a force less than F due to Y towards Y. The resultant of the first two forces is greater than F and is along the line PY away from Y. Since this is opposite in direction to the third force and is greater

in magnitude, the overall resultant of all 3 forces is along the line PY away from Y. [1]

3 **D**: equate the force expressions for Newton's law of gravitation and for Coulomb's law and make $G\varepsilon_0$ the subject to give $G\varepsilon_0 = Q_1Q_2/m_1m_2$ so the unit of $G\varepsilon_0$ is C^2/kg^2 [1]

4 **a** $V = kr^{-1}$ where k is constant. Using values for any point on the curve (e.g., $V = 2000\,V\,m$ at $r = 0.20\,m$) gives $k = Vr = 400\,V\,m$. [1]

Therefore at $r = 0.25\,m$, $V = k/r = 400\,V\,m/0.25\,m = 1600\,V$ [1]

b **i** $V = \dfrac{Q}{4\pi\varepsilon_0 r} = \dfrac{k}{r}$ therefore $\dfrac{Q}{4\pi\varepsilon_0} = k$ so

$Q = 4\pi\varepsilon_0 k = 4\pi \times 8.85 \times 10^{-12}\,F\,m^{-1} \times 400\,V\,m$ [1] $= 4.4(5) \times 10^{-8}\,C$ [1]

ii Electric field strength $E = \dfrac{Q}{4\pi\varepsilon_0 r^2} = \dfrac{V}{r}$

At $r = 0.25\,m$, $E = \dfrac{V}{r} = \dfrac{+1600\,V}{0.25\,m}$ [1] $= 6400\,V\,m^{-1}$ [1]

c **i** The conduction electrons in the rod are attracted towards the sphere and they therefore move towards end X [1] so end X becomes negatively charged because there are more conduction electrons there than (fixed) protons. [1].

The movement of the conduction electrons towards X means they move away from end Y [1] so there are more (fixed) protons at end Y than there are electrons so end Y becomes positively charged. [1]

ii Surface to X; the line is a curve from 2000 V to a lower potential at X than before [1]

X to Y; the line is flat (because the pd from X to Y is zero) [1] and at a potential between what the potential at X and at Y was. [1]

From Y; the potential decreases beyond from Y [1]

iii The electric field near P is stronger [1] The field will not be radial near P [1]

5 **a** **i** Resolving the tension T in the thread into horizontal and vertical components (see Figure 1) gives $F = T\sin\theta$ (horizontally) and $mg = T\cos\theta$ (vertically) [1]

Therefore $F/mg = \sin\theta/\cos\theta = \tan\theta$ so $F = mg\tan\theta$ [1]

▲ Figure 1

ii $F = mg\tan\theta = 0.82 \times 10^{-3}\,kg \times 9.8\,N\,kg^{-1} \times \tan 5.5° = 7.74 \times 10^{-4}\,N$ [1]

$E = V/d = 2800\,V/0.035\,m = 80\,000\,V\,m^{-1}$ [1]

Rearranging $F = QE$ gives $Q = \dfrac{F}{E}$

$= \dfrac{7.74 \times 10^{-4}\,N}{80\,000\,V\,m^{-1}}$ [1] $= 9.7 \times 10^{-9}\,C$ [1]

b θ decreases [1]; This is because the electric field strength $E = V/d$ so E decreases as V is unchanged and d increases [1].

Force F therefore decreases (as $F = QE$) and because $F = mg\tan\theta$, $\tan\theta$ decreases so θ decreases. [1]

6 **a** Let the charge of A $= Q_A$ and the charge of B $= Q_B$.

The potential of A at P, $V_A = \dfrac{1}{4\pi\varepsilon_0}\dfrac{Q_A}{r_1}$ where $r_1 = AP$

The potential of B at P, $V_B = \dfrac{1}{4\pi\varepsilon_0}\dfrac{Q_B}{r_2}$ where $r_2 = BP = 0.25\,m - 0.10\,m = 0.15\,m$

$V_A + V_B = 0$ at P therefore $\dfrac{1}{4\pi\varepsilon_0}\dfrac{Q_A}{r_1} + \dfrac{1}{4\pi\varepsilon_0}\dfrac{Q_B}{r_2} = 0$ [1]

So $\dfrac{Q_A}{r_1} = -\dfrac{Q_B}{r_2}$ so $Q_A = -\dfrac{Q_B}{r_2} \times r_1 = -0.10 \times \dfrac{Q_B}{0.15} = -2Q_B/3$ [1]

$Q_A + Q_B = +4.0\,nC$ so $(-2Q_B/3) + Q_B = +4.0\,nC$ or $Q_B/3 = +4.0\,nC$

Hence $Q_B = 12.0\,nC$, [1]

$Q_A = -2Q_B/3 = -8\,nC$ [1]

b **i** 3 lines of force including a straight line through P directed from B to A [1]

ii arc through P perpendicular to AB at P [1] bending towards [1] A on both sides of AB

Chapter 23

1 **B**: Time taken $= Q/I = CV/I$ which is B. [1]

2 **C**: In a charging circuit, the charge stored increases and the current decreases. So on this basis, C or D could be correct. However, the rate of change in both cases decreases so C is correct. [1]

3 **C**: The pd across the resistor decreases to $0.6V_B = (V_B - 0.40V_B)$ so the current decreases to 60% in time T. Therefore in time $2T$, the current and therefore the pd across the resistor decreases to 36% ($= 60\%$ of 60%) of V_B. So the pd across the capacitor in time $2T$ increases to 64% of the final pd which is $0.64V_B$. [1]

4 **a** **i** Q 0.90 mC 1.35 mC 1.80 mC 2.25 mC 2.70 mC [1]

ii Graph—correctly labelled axes [1]; linear scales from 0 covering at least half of each axis [1]; correctly plotted points and best-fit line [1]

iii Large gradient triangle drawn and correct read off to calculate the capacitance [1]
Capacitance calculated from graph to give $C = 0.90\,mF$ [1]

Answers to practice questions

b [Any 2 of the following 3 marks]

The initial rate of change of pd would be the same. [1]

The voltage would increase at a decreasing rate that becomes zero [1] when the pd is equal to the battery pd [1]

5 a Energy stored = ½ CV^2 = 0.5 × 2.2 × 10^{-3} F × $(5.5\,V)^2$ [1] = 3.3×10^{-2} J [1]

b i Assuming all the energy stored by the capacitor is transferred to the object's store of gravitational potential energy, $mg\Delta h = 3.3 \times 10^{-2}$ J [1]

Therefore 3.8×10^{-3} kg × 9.8 N kg^{-1} × $\Delta h = 3.3 \times 10^{-2}$ J [1] which gives $\Delta h = 3.3 \times 10^{-2}$ J/0.0373 N = 0.89 m [1]

ii Any 2 of the following 3 points:

Energy is dissipated (or transferred to the surroundings) by

1. the resistance heating effect of the discharge current [1]
2. friction between the moving parts in the motor [1]
3. sound waves created by the rotational motion (of vibrations) of the motor when it rotates [1]

6 a Rearranging $V = V_0 e^{-t/RC}$ gives $V_0/V = e^{t/RC}$ [1]

Taking natural logs on both sides gives ln V = ln V_0 − t/RC which is the equation for a straight line $y = mx + c$ [1] where the gradient $m = -1/RC$ and the y-intercept = ln V_0 [1]

b i Missing ln V values 1.281, 0.993, 0.765, 0.531, 0.300 [1]

Graph; axes labelled correctly [1] suitable scales covering at least half of each graph axis [1], points plotted correctly [1], best-fit line drawn [1] See Figure 2

▲ Figure 2

ii Large gradient triangle drawn [1] gradient correctly calculated from the height and base of the triangle [1]; gradient −1/RC (= −0.0083 s^{-1}) used to calculate C to give $C = 2.6$ mF [1] within ± 0.2 mF [1]

iii Time constant = RC = −1/gradient = 120 s (or correct calculation of RC) [1]

7 a i The dielectric molecules become polarised by the electric field so each molecule becomes a dipole with positive charge and negative charge at opposite ends of the molecule. [1] As a result, opposite surfaces on the dielectric perpendicular to the field become oppositely charged. [1] The surface nearest the negative plate of the capacitor is positively charged and the opposite surface positively charged. [1]

ii The positive charge that is created on the surface of the dielectric nearest the negative plate allows more electrons to transfer from the battery to the negative plate. [1] The negative charge that is created on the surface of the dielectric nearest the positive plate forces electrons to transfer to the battery from the positive plate. [1] As a result of this movement of electrons, more charge is stored by the capacitor as a result of the dielectric being inserted. [1]

b i The capacitance decreases. [1]

ii The charge stored Q is unchanged and the capacitance C decreases [1] so the potential difference increases (as it is equal to Q/C.) [1]

The energy stored increases [1] because it is equal to ½ QV and Q is unchanged and V increases. [1]

(Alternative for last mark: because work is done to overcome the attraction between the dielectric charge and the charge on each plate [1])

c Rearranging $C = \dfrac{A\varepsilon_0\varepsilon}{d}$

gives $d = \dfrac{A\varepsilon_0\varepsilon}{C} = \dfrac{1.2\,m^2 \times 8.85 \times 10^{-12}\,F\,m^{-1} \times 7}{47 \times 10^{-6}}$ [1]

= 1.6×10^{-6} m [1]

8 a Rearrange $v = V_0 e^{-\frac{t}{RC}}$ to give ln V = ln V_0 − t/RC [1]

Therefore t/RC = ln V_0 − ln V = ln 6.0 − ln 1.5 = 1.39 [1]

$t = RC \times 1.39 = 1.39 \times 50\,k\Omega \times 68\,\mu F = 4.7$ s [1]

b The time constant needs to be about 4 times longer (≈ 20 s/4.7 s) [1] so the resistor needs to be about 4 times bigger which is about 200 kΩ. [1]

Chapter 24

1 A: For no deflection, $BQv = QE$ therefore $E = Bv$ [1]

2 B: The k.e. of the particles = eV therefore ½ $mv^2 = eV$ hence its speed $v = (2eV/m)^{1/2}$. The magnetic force $Bev = mv^2/r$ therefore $r = mv/eB = m(2eV/m)^{1/2}/eB = (2mV/eB^2)^{1/2}$ [1]

3 C: X curves in the same direction as the proton so it also has a + charge which means B and D are incorrect. As the radius of curvature $r = mv/QB$ and Q is the same for a pion and a proton, r is proportional to the momentum. Since r is less for X, its momentum is less so A is incorrect. [1]

114

Answers to practice questions

4 **a** The magnetic force BQv on the particle is perpendicular to the direction of motion of the particle [1] so no work is done on the particle [1] and so its kinetic energy (and therefore its speed) is constant [1].

  b i The radius of curvature of the particle's path in the field, $r = 0.31\,\text{m}/2 = 0.155\,\text{m}$

   The mass of the ion = $20\,\text{u}$ = $20 \times 1.67 \times 10^{-27}\,\text{kg} = 3.34 \times 10^{-26}\,\text{kg}$ [1]

   $Bqv = mv^2/r$ hence $B = \dfrac{mv}{rq} =$

   $\dfrac{3.34 \times 10^{-26}\,\text{kg} \times 7.3 \times 10^{5}\,\text{m s}^{-1}}{0.155\,\text{m} \times 1.6 \times 10^{-19}\,\text{C}}$ [1] = 0.98 T [1]

  ii $r = mv/BQ$ therefore m is proportional to r [1]

   The ratio of distances from **P** = $0.34\,\text{m}/0.31\,\text{m}$ = 1.1 therefore the mass of the ions detected 0.34 m from **P** = $1.1 \times 20\,\text{u} = 22\,\text{u}$ [1]

  iii Each neon ion has 10 protons in its nucleus. The neon ions detected at 0.31 m from P each have 10 neutrons so their mass is 20 u [1] whereas the ions detected at 0.34 m each have 12 neutrons so their mass is 22 u. [1]

5 **a** Length of each turn $l = \pi d$ where d is the diameter of the coil.
   $F = BIln = 0.120\,\text{T} \times 0.19\,\text{A} \times \pi \times 0.082\,\text{m} \times 60$ [1] = 0.35 N [1]

  b i The force of the coil is perpendicular to the field. The a.c. repeatedly reverses its direction. [1] Each time it reverses, the force on the coil reverses so the direction of motion of the coil repeatedly reverses. [1]

  ii The magnitude of the force is proportional to the current in the coil. [1] If the rms (or peak) current is increased, the force increases in magnitude so the amplitude of vibration of the coil increases.[1]

6 **a** Weight of $W = 0.67 \times 10^{-3}\,\text{kg} \times 9.8\,\text{N kg}^{-1}$ = $6.57 \times 10^{-3}\,\text{N}$

   To regain balance, the moment due to the magnetic force = change of moment of W [1] = $6.57 \times 10^{-3}\,\text{N} \times 0.032\,\text{m} = 2.10 \times 10^{-4}\,\text{N m}$ [1]

   Force on AB = moment/distance from A to the pivot = $2.10 \times 10^{-4}\,\text{N m}/0.120\,\text{m}$ [1] = $1.75 \times 10^{-3}\,\text{N}$ [1]

  b Magnetic flux density $B = \dfrac{F}{Il} = \dfrac{1.75 \times 10^{-3}\,\text{N m}}{4.5\,\text{A} \times 0.095\,\text{m}}$ [1]

   $= 4.1 \times 10^{-3}\,\text{T}$ [1]

7 **a i** $F = BIln = 0.110\,\text{T} \times 0.67\,\text{A} \times 0.035\,\text{m} \times 160$ = 0.41(2) N [1]

  ii Torque = $Fd = 0.412\,\text{N} \times 0.028\,\text{m}$ [1] = $1.15 \times 10^{-2}\,\text{N m}$ [1]

  b The torque due to W = 1.4 N $\times$ (0.5 $\times$ 0.054 m) = 0.0378 N m [1].

   To raise the weight, an equal and opposite torque must be applied by the motor [1].

   The torque due the motor must be increased from 0.0115 N m to 0.0378 N m which is 3.3 $\times$ greater. [1]

   The torque is proportional to the motor current so the current must therefore be increased by $\times$ 3.3 to $3.3 \times 0.67\,\text{A} = 2.0\,\text{A}$ [1]

Chapter 25

1 **D:** The induced emf V = change of magnetic flux per second = BAn/t therefore $B = Vt/nA$. [1]

2 **C:** D is incorrect because V A^{-1} is the ohm. The unit of magnetic flux is the weber. The tesla equals 1 weber per square metre so A and B are incorrect. C is correct because induced emf = magnetic flux change per second therefore the unit of magnetic flux = the unit of emf $\times$ the unit of time = V s. [1]

3 **A:** The area per unit time swept out by the rod is constant and because the field is uniform, the flux per second swept out by the rod is constant so the induced emf is constant. [1]

4 **a** Faraday's law tells you that the induced emf = change of flux per second so the product of induced emf and time represents magnetic flux linkage. [1] Therefore, the area under the curve represents the total change of magnetic flux linkage [1]

  b No of small squares under the curve = 9 + 35 + 82 + 83 + 23 + 1 = 233 ± 5 [1]

   Each large square = 25 small squares represents 10 mV $\times$ 50 ms = 500 μWb = 0.50 mWb [1]

   Change of flux linkage = (233 ÷ 25) $\times$ 0.50 mWb = 4.7 mWb [1]

  c i Final flux linkage = 0 so initial flux linkage = 4.7 mWb [1]

   Area of coil $A = \pi \times (0.025\,\text{m})^2 = 1.96 \times 10^{-3}\,\text{m}^2$ [1]

   Magnetic flux density

   $B = \dfrac{\text{initial flux linkage}}{An} = \dfrac{0.0047\,\text{Wb}}{1.96 \times 10^{-3}\,\text{m}^2 \times 75}$

   = 0.032 T [1]

  ii % uncertainty in coil area = 2 $\times$ % uncertainty in diameter = (0.5/25) $\times$ 100% = 2.0% [1]

   % uncertainty in the magnetic flux estimate ≈ (5/233) $\times$ 100% = 2.1% [1]

   Overall % uncertainty = 4.1% so uncertainty in V = 4.1% $\times$ 4.7 mWb ≈ ±0.2 Wb [1]

5 **a i** Max flux linkage = BAn = 0.092 T $\times$ 2.3 $\times$ $10^{-3}\,\text{m}^2 \times 180$ [1] = 0.038(1) Wb [1]

  ii Peak emf = $BAn\omega = Ban(2\pi f) = 0.0381\,\text{Wb} \times 2\pi \times 50\,\text{Hz}$ [1] = 12 V [1]

  b correct shape 'sine wave or negative sine wave [1]; peak values at +12 V shown [1]; time period = 20 ms shown or correct time scale shown [1]

  c The peak-to-peak voltage would be increased to 1.8 its initial value (= initial value 90 Hz / 50 Hz) [1]

   The time period would be reduced to 0.56 its initial value (= initial value 50 Hz / 90 Hz) [1]

Answers to practice questions

6 a i $V_P = V_S \times \dfrac{N_P}{N_S} = 12\,V \times \dfrac{1800}{94}$ [1] $= 230\,V$ [1]

  ii Peak pd $= \sqrt{2} \times 230\,V = 325\,V$ [1]

b Efficiency $= \dfrac{I_S V_S}{I_P V_P} = \dfrac{50\,W}{0.23\,A \times 230\,V}$ [1] $= 0.95$
($= 95\%$) [1]

c Any 2 of the following:

Resistance heating [1] by the current in the coils [1]

Resistance heating [1] by eddy currents in the core [1]

Heating in the core due to [1] by repeated magnetisation and demagnetisation [1]

7 a i The conduction electrons in the disc are forced to move in/across the magnetic field [1] These electrons experience a magnetic force which is towards or away from the centre of the disc (depending on its direction of rotation) [1] If the electrons move to the centre, the centre becomes negative and the rim becomes positive (or if the electrons move to the rim, the rim becomes negative and the centre becomes negative) [1]

OR Each radial length in the disc cuts across the field lines [1] and sweeps out magnetic flux at a constant rate [1] so an emf is induced between the ends of each radial length with the ends near the centre have opposite polarity to the ends at the rim [1]

  ii Using the right hand rule; Rim is + and centre – [1]

b flux swept out per turn $= B\,A$ [1]

number of turns per second $= f$ therefore flux swept out per second $= BAf$ [1]

(magnitude of the) induced emf = flux swept out per second = BAf [1]

c Induced emf $= BAf = 0.270\,T \times \pi \times (0.015\,m)^2 \times 46\,Hz$ [1] $= 8.8\,mV$ [1]

8 Magnetic flux linkage $= BAn = 0.13\,T \times 0.018\,m \times 0.018\,m \times 55$ [1] $= 2.3 \times 10^{-3}\,Wb$ [1]

The magnetic flux linkage changes from $2.3\,mWb$ to $-2.3\,mWb$. Therefore the magnitude of the change of magnetic flux linkage $= (2.3\,mWb) - (-2.3\,mWb) = 4.6\,mWb$ [1]

The magnitude of the induced emf = change of magnetic flux linkage per second $= \dfrac{0.0046\,Wb}{0.092\,s}$ [1] $= 0.05\,V$ [1]

Chapter 26

1 D: The mass number decreases by 4 for each α decay and doesn't change in a β decay. Since the mass number decreases by 12, there must be 3 α decays. The proton number decreases by 2 for each α decay and increases by 1 in each β decay. Since the overall change of the proton number is a decrease of 2 and the α decays cause a decrease of 6, there must be 4 β decays to give an overall decrease of 2. Therefore there are 7 decays. [1]

2 B: For N_0 atoms initially, when there are N undecayed atoms, the number of decayed atoms is $3N$. Therefore $N_0 = N + 3N = 4N$ or $N/N_0 = ¼$

Time taken for N to decrease from N_0 to $¼\,N_0 = 2$ half-lives $= 20$ hours. [1]

3 B: Since $A = N\lambda$ and $\lambda = \ln 2/T^{1/2}$ then $N = A\,T^{1/2}/\ln 2$ so the ratio of the number of atoms of Q to P = their activity ratio × their half-life ratio

Half-life Q/P $= 100/50 = 2/1$

Activity of Q/P after 200 s $= (A_0/4)/(A_0/16) = 16/4 = 4/1$

Therefore ratio of number of atoms of P to Q = 8 [1]

4 a 11.3, 8.0, 6.2, 4.8, 3.8, 2.6, 2.1 [1]

b See Figure 1

Axes labelled, correctly plotted points [1]

Suitable scales [1]

Best-fit line [1]

▲ Figure 1 [3]

c i Large triangle clearly shown and used [1]

Gradient = –decay constant $\lambda = (2.40 - 1.00)/(150 - 0)\,s^{-1} = 0.00933\,s^{-1}$ [1]

Half life $= \ln 2/\lambda = 74\,s$ [1]

  ii The count rate changes if the distance from the source to the tube changes. [1] The distance needs to be constant so the decrease of the count rate is due only to radioactive decay [1]

5 a i Most of the mass of the atom is concentrated in its nucleus which has a much smaller diameter than the atom [1] The nucleus is + charged and it repels and scatters α particles that come close to it [1] Only a small proportion of the incident α particles come close enough to be scattered by more than 90° [1]

  ii The probability of an α particle being deflected by a nucleus is very small because most of the atom is empty space. [1] The foil has multiple layers of atoms and the probability of an α particle between scattered more than once is very very small [1] as it is equal to the probability of being scattered once raised to the power n where $n > 1$ [1]

b $E_K = \dfrac{Q_\alpha Q_N}{4\pi\varepsilon_0 d}$ gives $d = \dfrac{Q_\alpha Q_N}{4\pi\varepsilon_0 E_K}$

$= \dfrac{2e \times 29e}{4\pi \times 8.85 \times 10^{-12}\,\text{F m}^{-1} \times 4.5 \times 10^6\, e}$ [1]

$= \dfrac{58 \times 1.6 \times 10^{-19}\,\text{C}}{4\pi \times 8.85 \times 10^{-12}\,\text{F m}^{-1} \times 4.5 \times 10^6}$ [1]

$= 1.9 \times 10^{-14}\,\text{m}$ [1]

6 a Rearranging $C = \dfrac{k}{(d + d_0)^2}$ gives $d + d_0 = \dfrac{\sqrt{k}}{\sqrt{C}}$ [1]

Therefore $d = \dfrac{a}{\sqrt{C}} + b$ where $b = -d_0$ and $a = \sqrt{k}$ [1]

b i Values of $1/\sqrt{C}$ correct to 3 sf [1]

Units correct [1]

d/mm	102	136	170	202	245	305	385
C/s⁻¹	218	126	73	57	40	27	15
$1/\sqrt{C}$/s$^{1/2}$	0.0678	0.0891	0.117	0.132	0.158	0.192	0.258

ii See Figure 2

Points plotted correctly and axes labelled correctly [1]

Suitable scales [1]

Best fit line [1]

▲ Figure 8

c $d_0 = 4$ to 5 mm [1]

7 a $^{63}_{29}\text{Cu} + ^{1}_{0}\text{n} \rightarrow ^{64}_{29}\text{Cu}$ [1]

b i Beta minus radiation [1]: The copper-64 nucleus is neutron-rich [1] and becomes stable by emitting a β⁻ particle [1]

ii $^{64}_{29}\text{Cu} \rightarrow ^{64}_{30}\text{Zn}$ [1] $+ ^{0}_{-1}\beta + \bar{v}$ [1]

c If the copper-63 nucleus absorbs a proton, it would become a zinc-64 nucleus [1] which is stable. [1]

8 a The radiation from the source included α particles with a range just less than the tube–source distance at which the count rate dropped. [1]

These particles were unable to reach the tube at this distance. [1]

b The source must be emitting either longer-range α radiation and/or β radiation and/or γ radiation. [1]

Placing paper between the source and the tube would cause a further drop in the count rate if the source was emitting longer-range α radiation. [1]

Placing a 5 mm thick aluminium plate instead would cause a drop in the count rate if the source was emitting β radiation. [1] If no drop occurred, the source must be emitting γ radiation. [1]

Chapter 27

1 D: The mass defect $= Z m_p + (A - Z) m_n - M = m_n A + (m_p - m_n) Z - M$ therefore $r = m_n$, $s = (m_p - m_n)$ and $t = -1$ so D is correct. [1]

2 B: Fission energy/second = 1200 MW/0.4 = 3000 MW so number of fission events per second = 3000 MW/(200 MeV × 1.6 × 10⁻¹³ J/MeV) = 9.38 × 10¹⁹ ≈ 10²⁰ [1]

3 D: The total number of protons is unchanged at 92 so Y has 92 – Q protons which means A and B are incorrect. The total number of protons and neutrons decreases by 1 from 235 as 2 fission neutrons are released after a neutron has been absorbed. So Y has 234 – P protons and neutrons. [1]

4 a $^{64}_{29}\text{Cu} \rightarrow ^{0}_{+1}\beta + ^{64}_{28}\text{Ni}$ [1] $+ \bar{v}$ [1] $+ Q$

b mass difference $\Delta m = 0.66\,\text{MeV}/931.5\,\text{MeV u}^{-1} = 0.00071\,\text{u}$ [1]

mass of nickel nucleus = mass of copper nucleus – mass of positron – Δm [1]

$= 63.91384 - 0.00055 - 0.00071\,\text{u} = 63.91258\,\text{u}$ [1]

c a = up quark (u); b = down quark (d) [1]
c = a W⁺ boson; d = an (electron) neutrino (v_e) [1]

d i similarity: both are elementary (or leptons) [1]

difference: neutrino is uncharged whereas the β⁺ particle is charged (or the neutrino is a matter particle whereas the β⁺ particle is an antimatter particle) [1]

ii similarity: both are charged [1]

difference: the W⁺ boson is very short-lived compared with the β⁺ particle [1]

5 a i See Topic 27.2, Figure 2: Curve showing an increase up to about $A = 55$ [1] then a more gradual decrease up $A \approx 240$ [1] at BE/nucleon > 6 MeV [1]

Maximum BE/nucleon between 8 to 9 MeV [1]

ii The average binding energy per nucleon increases in fusion [1] provided the atomic mass of the fused nucleus is less than about 55. [1] The number of nucleons is unchanged [1] so the total binding energy increases [1]

Answers to practice questions

b i Mass defect $\Delta m = 2m_p + 2m_n - M_{NUC} = 0.03040\,u$ [1]

Therefore, binding energy = $0.03040\,u \times 931.5\,MeV/u = 28.3\,MeV$ [1]

Binding energy per nucleon = $28.3\,MeV/4 = 7.1\,MeV$ [1]

ii Correct position at $A = 4$ [1]

iii The energy released when protons and neutrons form a cluster (or a nucleus) is equal to the binding energy of the cluster. [1]

So the energy released when a proton and a neutron form a ^2_1H nucleus is much less than the energy released when an alpha particle forms from 2 protons and 2 neutrons. [1]

A ^2_1H nucleus formed inside a large nucleus does not gain sufficient energy to leave the large nucleus whereas an alpha particle does gain sufficient energy to leave. [1].

6 a Induced fission is the splitting of a nucleus of a fissile material into two approximately equal smaller nuclei [1] when a neutron from outside the original nucleus collides with it. [1]

b i When a neutron causes a nucleus to undergo fission, 2 or 3 neutrons are released in the process. [1] These neutrons can go on to produce further fission events in which more neutrons are released that cause further fission events. [1] In this way, a chain of reactions occurs until all the nuclei of the fissile material have undergone fission. [1]

A minimum mass (the critical mass) is required because if the mass is too low, too many of the fission neutrons escape from the fissile material without causing fission [1] and some are absorbed by other nuclei without fission. [1]

ii For a chain reaction to occur, at least one fission neutron per fission on average must go on to produce further fission. [1] The rate of fission events is proportional to the mass of the material. [1] A minimum mass (the critical mass) is required because if the mass is too low, too many of the fission neutrons escape from the fissile material without causing fission [1] as the surface area in relation to the mass is too great. [1]

c i The control rods are used to absorb excess neutrons to ensure a constant rate of fission events occurs. [1]. The depth of the rods in the reactor core is controlled automatically according to the rate at which fission events occur. [1] If the rate increases, the rods are pushed further into the reactor core so they absorb more neutrons, causing the rate of fission events to decrease to where it was. [1] If the rate of fission events decreases, the rods are withdrawn gradually to increase the rate of fission events back to where it was. [1]

Either from: boron or cadmium [1]

ii Fission neutrons are released with too much speed to cause further fission of uranium 235 nuclei [1] and their kinetic energy must be reduced considerably before they can cause induced fission. [1] Their kinetic energy is reduced by allowing them to undergo repeated elastic collisions with the atoms of the moderator [1] as some of their kinetic energy is transferred to the moderator atoms in each collision. [1] The amount of kinetic energy transferred in such a collision is greater the closer the mass of the moderator atoms is to the mass of a neutron. [1]

iii Any 2 points: The coolant must be a liquid or a gas that can be pumped through channels in the moderator.[1] It must have a high specific heat capacity [1] and not react chemically with the materials in the reactor core. [1]

The coolant is pumped through pipes from the reactor core to a heat exchanger [1] where it is used to heat water and create steam to drive turbines [1]. After passing through the heat exchanger, the coolant returns through pipes to the reactor core where it is heated up again before returning to the heat exchanger. [1]

7 a i $^{210}_{84}Po \rightarrow\ ^{206}_{82}Pb$ [1] $+\ ^4_2\alpha$ [1] $+ Q$

ii mass difference $\Delta m = 209.93675\,u - 205.92945\,u - 4.00150\,u = 0.00580\,u$ [1]

$Q = 931.5\,MeV/u \times 0.00580\,u$ [1] $= 5.4\,MeV$ [1]

b i The alpha particle and the lead nucleus move away from each other with equal and opposite momentum [1]

Therefore the energy released is shared between them as kinetic energy so the kinetic energy of the alpha particle is less than the energy released. [1]

ii As $E_K = p^2/2m$ and the momentum p is equal in magnitude for the alpha particle and the lead nucleus [1], the ratio of their kinetic energies is the inverse of their mass ratio [1]

Their mass ratio $m_{Pb}/m_\alpha = 206/4 = 51.5$ [1]

Therefore the $E_{KPb}/E_{K\alpha} = m_\alpha/m_{Pb} = 1/51.5$ so $E_{KPb} = E_{K\alpha}/51.5 = 0.0194\,E_{K\alpha}$ [1]

Since $E_{K\alpha} + E_{Pb} = Q = 5.4\,MeV$ then $E_{K\alpha} + 0.0194\,E_{K\alpha} = 5.4\,MeV$ [1]

Therefore $E_{K\alpha} = (5.4/1.0194)\,MeV = 5.3\,MeV$ [1]

Answers to summary questions

17.1

1. **a** Circumference of wheel = π × diameter
   = π × 0.60 m = 1.885 m
   Number of turns = $\frac{\text{distance}}{\text{circumference}} = \frac{100\,\text{m}}{1.885\,\text{m}}$
   = 53.1 = 53 (2 s.f.) [1]

   b Time taken to travel 100 m = 100 m/25 m s^{-1} = 4.0 s
   Frequency f = number of turns/time taken = 53.1/4.0 s = 13.3 Hz = 13 Hz (2 s.f.) [1]

   c Angular speed $\omega = 2\pi f = 2\pi \times 13.3$ Hz = 83.6 rad s^{-1} = 84 rad s^{-1} (2 s.f.) [1]

2. **a** Frequency $f = \frac{1}{\text{time for 1 rotation}} = \frac{1}{(93 \times 60\,\text{s})}$
   = 1.8 × 10^{-4} Hz [1]

   b Speed $v = \omega r = 2\pi f r = 2\pi \times 1.8 \times 10^{-4}$ Hz × (6400 + 400) × 10^3 m ≈ 7.7 km s^{-1} [1]

3. **a** Angular speed $\omega = 2\pi f = 2\pi \times 0.95$ Hz = 6.0 rad s^{-1} [1]

   b The two gear wheels move at the same speed, v, where they are in contact. [1]
   Therefore $v = \omega_1 r_1 = \omega_2 r_2$ so $\omega_2 \times 2.5$ mm = 6.0 rad s^{-1} × 8.2 mm [1]
   Hence ω_2 = (6.0 rad s^{-1} × 8.2)/2.5 mm = 19.7 rad s^{-1}
   Rearranging $\omega = 2\pi f$ gives $f = \frac{\omega}{2\pi} = \frac{19.7\,\text{rad s}^{-1}}{2\pi}$
   = 3.1 Hz [1]

   c i $\Delta\theta = \omega \Delta t$ = 19.7 rad s^{-1} × 0.02 s = 0.39(4) rad [1]

   ii speed of tooth, $v = \omega_2 r_2$ = 19.7 rad s^{-1} × 2.5 mm = 49 mm s^{-1}
   distance moved $\Delta s = v \Delta t$ = 49 mm s^{-1} × 0.02 s
   = 0.98 mm
   (or $\Delta s = r \Delta\theta$ = 2.5 mm × 0.39(4) rad
   = 0.98 mm) [1]

17.2

1. **a** Angular speed $\omega = 2\pi/T = 2\pi/2.3$ s = 2.7(3) rad s^{-1} [1]

   b Speed $v = \omega r$ = 2.73 rad s^{-1} × 1.1 m = 3.0 m s^{-1} [1]

   c Centripetal force = $m \omega^2 r$ = 28 kg × (2.73 rad s^{-1})^2 × 1.1 m = 230 N [1]

2. Centripetal force = $\frac{mv^2}{r} = \frac{22\,\text{kg} \times (0.78\,\text{ms}^{-1})^2}{2.8\,\text{m}}$ = 4.8 N [1]

3. **a** Angular speed = $2\pi/T = 2\pi/12$ s = 0.52(4) rad s^{-1}
   Speed $v = \omega r$ = 0.524 rad s^{-1} × 2.5 m = 1.3(1) m s^{-1} [1]

   b Centripetal force = $m \omega^2 r$ = 61 kg × (0.524 rad s^{-1})^2 × 2.5 m = 42 N [1]

17.3/17.4

1. **a** Centripetal acceleration $a = \frac{v^2}{r} = \frac{(22\,\text{ms}^{-1})^2}{82\,\text{m}}$
   = 5.9 m s^{-2} [1]

   b Support force $S = mg + ma$ (as $S - mg = ma$) [1]
   Therefore S = (1200 kg × 9.81 m s^{-2}) + (1200 kg × 5.9 m s^{-2}) = 1.9 × 10^4 N [1]

2. Angular speed $\omega = 2\pi/T = 2\pi/7.0$ s = 0.90 rad s^{-1} [1]
   For a passenger at the top, the centripetal acceleration = $\omega^2 r$ = (0.90 rad s^{-1})^2 × (16 m)
   = 13.0 m s^{-2}
   Support force $S = ma - mg$ (as $S + mg = ma$) [1]
   Therefore S = (48 kg × 13.0 m s^{-2}) − (48 kg × 9.81 m s^{-2}) = 153 N = 150 N (2 s.f.) [2]

3. Centripetal acceleration $a = \frac{v^2}{r} = \frac{(2.8\,\text{ms}^{-1})^2}{3.3\,\text{m}}$
   = 2.38 m s^{-2} [1]
   Support force $S = mg + ma$ (as $S - mg = ma$)
   Therefore S = (23 kg × 9.81 m s^{-2}) + (23 kg × 2.38 m s^{-2}) = 280 N [2]

18.1/2

1. **a i** 0.771 s [1]
   ii 1.30 Hz [1]

   b Range of timings = ±0.2 s, mean timing = 15.4 s [1]
   % uncertainty in frequency = % uncertainty in timings = range of timings/mean value ×100% [1] = (±0.2 s/15.4 s) × 100% = 1.3% [1]

2. **a** $\omega = 2\pi/T = 2\pi/0.77$ s = 8.2 rad s^{-1} [1]
   acceleration $a = -\omega^2 x = -(8.2\,\text{rad s}^{-1})^2 \times 0.025$ m
   = −1.7 m s^{-2} [1]

   b $x = 0$ so $a = -\omega^2 x = 0$ [1]

   c As $x = -25$ mm, from part **a** $a = (-\omega^2 x$ = −(8.2 rad s^{-1})^2 × −0.025 m) = +1.7 m s^{-2} [1]

18.3

1. **a** Angular frequency $\omega = 2\pi/T = 2\pi/2.3$ s = 2.73 rad s^{-1} = 2.7 rad s^{-1} (2 s.f.) [1]

   b Rearranging $a = -\omega^2 x$ gives $x = -a/\omega^2$ therefore at maximum displacement, $x = +A$ when $a = -7.2$ m s^{-2}, $-A = -7.2$ m s^{-2}/(2.73 rad s^{-1})^2 [1]
   = −0.97 m. Therefore A = 0.97 m. [1]

2. **a** Sketch graph as SB 18.3 Figure 3 top graph; sine wave [1]
   time for 1 cycle shown as 2.3 s [1]

   b $x = A \sin(\omega t)$ = 0.97 m × sin (2.73 rad s^{-1} × 2.0 s)
   = −0.71 m [1]

3. **a** After one half cycle, its displacement $x = -55$ m (because it is at the opposite extreme to its initial displacement) [1]
   Its angular frequency $\omega = 2\pi/T = 2\pi/0.27$ s = 23.3 rad s^{-1} [1]
   Therefore at $x = -55$ mm, its acceleration = $-\omega^2 x$
   = −(23.3 rad s^{-1})^2 × −0.055 m = 30 m s^{-2} [1]
   Therefore the magnitude of its acceleration = 30 m s^{-2} [1]

Answers to summary questions

b At time $t = 0.20$ s, its displacement $x = A \cos \omega t$
   $= 55$ mm $\times \cos(23.3$ rad s$^{-1} \times 0.20$ s$) = -2.9$ mm [1]

18.4

1 a i Extension of spring at equilibrium is $\Delta L_0 = 79$ mm $= 0.079$ m [1]

   ii Spring constant $k = \dfrac{mg}{L_0} = \dfrac{0.20 \times 9.8}{0.079} = 25$ N m^{-1} [1]

b $T = 2\pi \sqrt{\dfrac{0.20 \text{ kg}}{25 \text{ N m}^{-1}}} = 0.56$ s [1]

2 a From the equation $a = -250x$, $\omega_2^2 = 250$ rad^2 s^{-2} [1]
   so $T = 2\pi/(250 \text{ rad}^2 \text{ s}^{-2})^{\frac{1}{2}} = 0.40$ s [1]

b From the equation $T = 2\pi \sqrt{m/k}$, $T \propto m^{\frac{1}{2}}$ so $m \propto T^2$ [1]
   therefore $m_2/m_1 = T_2^2/T_1^2 = (0.40/0.56)^2 = 0.51$ [1]
   which gives $m_2 = 0.51 m_1 = 0.51 \times 0.20$ kg $= 0.10$ kg [1]

3 a $T = 2\pi \sqrt{\dfrac{L}{g}} = 2\pi \sqrt{\dfrac{0.250 \text{ m}}{9.81 \text{ m s}^{-2}}} = 1.00(3)$ s [1]

b For $L = 255$ mm, $T = 2\pi \sqrt{\dfrac{L}{g}} = 2\pi \sqrt{\dfrac{0.255 \text{ m}}{9.81 \text{ m s}^{-2}}} = 1.01(3)$ s [1]
   Therefore the uncertainty in $T = \pm 0.01$ s which is $\pm 1\%$ of T [1]

   Alternative method
   % uncertainty in $L = \pm(\Delta L/L) \times 100\% = \pm(5$ mm/250 mm$) \times 100\% = \pm 2\%$ [1]
   As $T \propto L^{\frac{1}{2}}$, % uncertainty in $T = \frac{1}{2} \times$ % uncertainty in $L = 0.5 \times \pm 2\% = \pm 1\%$ [1]

18.5

1 a $T = 2.2$ s so angular frequency $\omega = 2\pi/T = 2\pi/2.2$ s $= 2.86$ rad s^{-1} [1]

b Rearranging $T = 2\pi \sqrt{m/k}$ gives $k = 4\pi^2 (m/T^2)$ [1]
   Therefore $k = 4\pi^2 (0.150$ kg$/(2.2)^2) = 1.22$ N m^{-1} [1]

c Amplitude $A = 30$ mm so maximum speed $= \omega A = 2.86$ rad s$^{-1} \times 0.030$ m $= 0.09$ m s^{-1} [1]

2 a Total energy $= \frac{1}{2} k A^2 = 0.5 \times 1.22$ N m s$^{-1} \times (0.030$ m$)^2 = 0.55$ mJ [1]

b PE curve parabola from origin [1]
   PE at $\pm A =$. max PE [1]
   KE curve inverted PE curve [1] with max KE (= min PE) at $x = 0$ and zero KE at $\pm A$ [1]

3 a Its maximum speed $v_{\max}$ is given by $\frac{1}{2} m v_{\max}^2 = mgh$ where $h = 10.0$ mm [1]
   Therefore $v_{\max} = (2gh)^{\frac{1}{2}} = 0.44(3)$ m s^{-1} [1]

b i $T = 1.50$ s so angular frequency $\omega = 2\pi/T = 2\pi/1.50$ s $= 4.19$ rad s^{-1} [1]
   Maximum speed $= \omega A$ therefore
   $A =$ Maximum speed/ω
   $= 0.44(3)$ m s$^{-1}/4.19$ rad s$^{-1} = 0.106$ m [1]

ii At $x = 0.5 A$, speed
   $v = \omega \sqrt{A^2 - x^2} = 4.19 \sqrt{0.106^2 - 0.053^2}$ m
   $= 0.385$ m s^{-1} [1]

18.6

1 a The object and the spring are in resonance at this frequency [1]
   because the periodic force is in phase with the velocity of the object at this frequency [1]

b The object gains energy from the work done by the periodic force [1]
   and its amplitude increases until the resistive forces due to damping becomes equal and opposite to the periodic force. [1]

2 a Because the time period T is proportional to the square root of the object's mass, increasing the mass by a factor of 4 increases the time period by $\times 2$ $(= \sqrt{4})$ [1]
   So the natural frequency of oscillation is halved and resonance occurs at $0.5 f_0$ [1]

b If a stiffer spring is used, the spring constant k is larger therefore the time period T is reduced as T is proportional to $1/\sqrt{k}$. [1]
   Therefore the natural frequency of the system is increased so resonance takes place at a higher frequency [1]

3 a Rearranging $T = 2\pi \sqrt{m/k}$ gives $m = kT^2/4\pi^2$ where $T = 1/f = 1/1.8$ Hz $= 0.56$ s [1]
   Therefore $m = 25$ N m$^{-1} \times (0.56$ s$)^2/4\pi^2$ [1] $= 0.20$ kg [1]

b Using two springs in parallel doubles the spring constant to 50 N m^{-1} (see Topic 11.2). [1]
   As T is proportional to $1/\sqrt{k}$ and f is proportional to $1/T$, then f is proportional to $\sqrt{k}$. [1]
   As k is doubled, the frequency of oscillation is increased by $\times \sqrt{2}$ to $1.8\sqrt{2}$ Hz which equals $2.5(5)$ Hz [1]

19.1

1 a The internal energy of an object is the sum of the kinetic and potential energies of its molecules' positions due to their random motion and positions. [1]

b Its internal energy increases by 100 J $(= 400$ J gained due to work done $- 300$ J lost due to energy transfer from it by heating) [1]

2 a The average kinetic energy decreases and the average potential energy is unchanged. [1]

b The average potential energy decreases and the average kinetic energy is unchanged as the substance transfers energy by heating to its surroundings. [1]

Answers to summary questions

3 When the rubber band is released, the potential energy of its molecules decreases as they move closer together due to the forces (or bonds) between them [1]. These forces pull the molecules together and they collide with each other and vibrate more [1]. So the work done by these forces reduces their potential energy and increases their kinetic energy and the temperature of the rubber increases [1]. Energy transfer to the surroundings takes place and the rubber band loses internal energy and cools down to the same temperature as its surroundings. [1]

19.2

1 Energy needed to heat the tank = $mc\,\Delta T$ = 20 kg × 390 J kg^{-1} K^{-1} × (60°C − 15°C) = 3.51 × 10^5 J [1]

Energy needed to heat the tank = $mc\,\Delta T$ = 430 kg × 4200 J kg^{-1} K^{-1} × (60°C − 15°C) = 8.127 × 10^7 J

Total energy needed = 8.16 × 10^7 J = 8.2 × 10^7 J (2 s.f.) [1]

2 Energy to heat the aluminium = PE loss therefore $mc\Delta T = m\,g\,h$ [1]

Therefore temperature increase $\Delta\theta = g\,h/c$ = (9.81 m s^{-1} × 8000 m)/900 J kg^{-1} K^{-1} = 87 K [1]

3 Mass flow in 1 second, m = density × volume flow in 1 second = 1000 kg m^{-3} × 3.8 × 10^{-5} m^3 = 0.038 kg [1]

Energy supplied in 1 second E = 5000 J = $m\,c\,\Delta T$ [1]

therefore $\Delta T = E/m\,c$ = 5000 J / (0.038 kg × 4200 J kg^{-1} K^{-1}) = 31 K [1]

19.3

1 When a liquid cools, the molecules move slower as the temperature decreases [1] so the force bonds are able to pull the molecules closer and closer together [1]. When the temperature decreases to the melting point, the molecules hold each other in fixed positions and the liquid solidifies at its melting point. [1]

2 Energy transferred to cool the water to 0°C = $mc\,\Delta T$ = 0.095 kg × 4200 J kg^{-1} K^{-1} × 12 K = 4790 J [1]

Energy transferred to freeze the water = mL = 0.095 kg × 3.40 × 10^5 J kg^{-1} = 32300 J [1]

Energy transferred per second = (32300 J + 4790 J)/600 s = 62 J s^{-1} [1]

3 a Energy used to boil 0.015 kg of water away = mL = 0.015 kg × 2.25 × 10^6 J kg^{-1} = 33750 J [1]

Energy supplied by kettle = 3000 W × 20 s = 60000 J

% energy supplied that is used to boil the water = 33750 J/60000 J × 100% = 56% [1]

b Energy gained by the water in the beaker = $m_w c\,\Delta T$ = 0.11 kg × 4200 J kg^{-1} K^{-1} × (55°C − 20°C) = 1.62 × 10^4 J = 1.6 × 10^4 J (2 s.f.) [1]

Energy transferred to the water when the steam condensed at 100°C = $mL = m \times 2.25 \times 10^6$ J kg^{-1} where m is the mass of the condensed steam [1]

Energy transferred to the water when the condensed steam cooled from 100°C to 55°C = $mc\,\Delta T = m \times 4200$ J kg^{-1} K^{-1} × (100°C − 55°C) = $m \times 1.89 \times 10^5$ J kg^{-1} [1]

Assuming no energy is transferred to the surroundings,

energy gained

by the water = energy transferred from the steam and the condensed steam

Therefore $(m \times 2.25 \times 10^6$ J kg$^{-1}) + (m \times 1.89 \times 10^5$ J kg$^{-1}) = 1.62 \times 10^4$ J [1]

$(2.25 \times 10^6$ J kg$^{-1} + 1.89 \times 10^5$ J kg$^{-1}) \times m = 1.62 \times 10^4$ J

2.44×10^6 J kg$^{-1} \times m = 1.62 \times 10^4$ J

$m = 1.62 \times 10^4$ J / 2.44×10^6 J kg^{-1} = 6.6 × 10^{-3} kg [1]

20.1

1 Stage 1; Use $p_2 V_2 = p_1 V_1$ to give, after compression, pressure $p_2 = p_1 V_1/V_2$ = 120 kPa × 8.0 × 10^{-4} m^3 / 5.0 × 10^{-4} m^3 = 192 kPa [1]

Stage 2; Use $p_2/T_2 = p_1/T_1$ to give, after compression, [1]

pressure $p_2 = p_1 T_2/T_1$ = 192 kPa × (20 + 273) K / (50 + 273) K = 174 kPa = 170 kPa (2 s.f.) [1]

2 a Pressure $p_2 = p_1 T_2/T_1$ = 100 kPa × 173 K / 373 K = 46 kPa [2]

b Graph; straight line with a +gradient; line shown between 173 and 373 K with one temp and pressure values shown correctly; correct gradient or 2nd values shown correctly for either 173 or 373 K pressure value [3]

3 New volume $V_2 = V_1 T_2/T_1$ = 0.085 m^3 × 328 K / 288 K = 0.097 m^3 [2]

20.2

1 a Rearrange $pV = nRT$ to give $n = pV/RT$ = 120 kPa × 0.020 m^3/8.31 J mol^{-1} K^{-1} × (273 + 20) = 0.99 mol [1]

Mass = 0.99 mol × 0.028 kg = 0.028 kg [1]

b Pressure $p_2 = p_1 T_2/T_1$ = 120 kPa × 278 K / 293 K = 114 kPa [1]

2 Rearrange $pV = nRT$ to give $n = pV/RT$ = 101 kPa × 2.2 × 10^{-5} m^3/8.31 J mol^{-1} K^{-1} × (273 + 12) = 9.38 × 10^{-4} mol [1]

Mass = 9.38 × 10^{-4} mol × 0.032 kg mol^{-1} = 3.0 × 10^{-5} kg [1]

3 a Rearrange $pV = nRT$ for 1.0 m^3 to give $n = pV/RT$ = 100 kPa × 1.0 m^3/8.31 J mol^{-1} K^{-1} × 300 K = 40.1 mol [1]

Number of molecules in n moles = $n N_A$ = 40.1 mol × 6.02 × 10^{23} mol^{-1} = 2.41 × 10^{25} [1]

Volume per molecule = 1/2.41 × 10^{25} = 4.1 × 10^{-26} m^3 [1]

b Assume each molecule occupies a cube of length d so d^3 = 4.1 × 10^{-26} m^3 [1]

The average spacing d = (4.1 × 10^{-26} m^3)$^{1/3}$ = 3.4 × 10^{-9} m. [1]

Answers to summary questions

20.3

1 The pressure increases because reducing its volume means the gas molecules travel less distance on average between impacts at the walls. [1]

Therefore, there are more impacts per second so the pressure is greater. [1]

2 Mass of a hydrogen molecule =
$0.002 \text{ kg mol}^{-1}/6.02 \times 10^{23} \text{ mol}^{-1} = 3.32 \times 10^{-27} \text{ kg}$ [1]
Mean kinetic energy $\frac{1}{2} mc_{rms}^2 = \frac{3}{2} kT = 1.5 \times 1.38 \times 10^{-23} \text{ J K}^{-1} \times 293 \text{ K} = 6.07 \times 10^{-21} \text{ J}$ [1]

Therefore rms speed = $(2 \times \text{mean KE}/m)^{1/2}$ = $(2 \times 6.07 \times 10^{-21} \text{ J}/3.32 \times 10^{-27} \text{ kg})^{1/2} = 1910 \text{ m s}^{-1} = 2000 \text{ m s}^{-1}$ (1 s.f.) [1]

3 a Rearrange $pV = nRT$ for 1.0 m^3 to give $n = pV/RT$ = $120 \text{ kPa} \times 0.018 \text{ m}^3/8.31 \text{ J mol}^{-1} \text{K}^{-1} \times 323 \text{ K}$ = $0.805 \text{ mol} = 0.81 \text{ mol to (2 s.f.)}$ [1]

b Internal energy = $\frac{3}{2} nRT$ =
$1.5 \times 0.805 \times 8.31 \text{ J mol}^{-1} \text{K}^{-1} \times 323 \text{ K} = 3.2 \times 10^3 \text{ J}$ [2]

4 The mean KE is the same for the CO_2 molecules and the nitrogen molecules (as they are at the same temperature). [1]

Therefore, $\frac{1}{2} mc_{rms}^2$ for CO_2 molecules = $\frac{1}{2} mc_{rms}^2$ for nitrogen molecules [1]

$$\frac{c_{rms} \text{ for } CO_2}{c_{rms} \text{ for nitrogen}} = \sqrt{\frac{\text{mass of a nitrogen molecule}}{\text{mass of a } CO_2 \text{ molecule}}}$$

$$= \sqrt{\frac{\text{molar mass of nitrogen}}{\text{molar mass of } CO_2}}$$

$$= \sqrt{\frac{0.044 \text{ kg}}{0.028 \text{ kg}}} = 1.25 \quad [2]$$

5 a Rearrange $pV = nRT$ to give $n = pV/RT = 1.00 \times 10^5 \text{ Pa} \times 0.037 \text{ m}^3/(8.31 \text{ J mol}^{-1} \text{K}^{-1} \times 300 \text{ K})$ = $1.48 \text{ mol} = 1.5 \text{ mol}$ (2 s.f.) [1]

b For 1 mole, $pV = \frac{1}{3} Nm c_{rms}^2 = \frac{1}{3} m c_{rms}^2$ where M = molar mass [1]

So $c_{rms}^2 = 3 pV/M = 3 \times 1.00 \times 10^5 \text{ Pa} \times 0.037 \text{ m}^3/0.032 \text{ kg} = 3.47 \times 10^5 \text{ m}^2 \text{ s}^{-2}$ [1]
Therefore $c_{rms} = 590 \text{ m s}^{-1}$ [1]

c Mass of gas = $1.48 \text{ mol} \times 0.032 \text{ kg mol}^{-1} = 0.0474 \text{ kg}$ [1]

Density = mass/volume = $0.0474 \text{ kg}/0.037 \text{ m}^3 = 1.28 \text{ kg m}^{-3} = 1.3 \text{ kg m}^{-1}$ (2 s.f.) [1]

6 Rearrange $pV = nRT$ for $V = 1.0 \text{ m}^3$ to give $n = pV/RT$ = $1.01 \times 10^5 \text{ Pa} \times 1.0 \text{ m}^3/(8.31 \text{ J mol}^{-1} \text{K}^{-1} \times 288 \text{ K})$ = 42.2 mol. [1]

Mass of gas = $42.2 \text{ mol} \times 0.029 \text{ kg mol}^{-1} = 1.22 \text{ kg}$ [1]

Density = mass/volume = $1.22 \text{ kg}/1.0 \text{ m}^3 = 1.2(2) \text{ kg m}^{-3}$ [1]

21.1

1 a 1.6 N kg^{-1} [1]

b 400 N [1]

2 a In a uniform gravitational field, the gravitational field has the same magnitude and direction at all points in the field [1]

b 3.5 N [1]

3 Weight on Mars = $(650 \text{ N}/9.81 \text{ N kg}^{-1}) \times 3.70 \text{ N kg}^{-1}$ [1] = 245 N [1]

21.2

1 a 7.4 KJ [1]

b 490 J kg^{-1} [1]

2 a The gravitational potential at a point is the work done per unit mass to move a small object from infinity to that point. [1]

b Work done $W = mgh$ therefore $\Delta V = W/m = gh$ [1]
= $9.81 \text{ N kg}^{-1} \times 20.0 \text{ m} = 196 \text{ J kg}^{-1}$ [1]

3 a For $\Delta x = +500$ km (+ for upwards),
$\Delta V = +4.6 \text{ MJ kg}^{-1}$ [1]
$g = -\Delta V/\Delta x = -(4.6 \text{ MJ kg}^{-1})/500$ km [1] = $(-)9.2 \text{ N kg}^{-1}$ [1]

b $\Delta E_p = m\Delta V = 300 \text{ kg} \times (+)4.6 \text{ MJ kg}^{-1}$ [1] = $1.38 \times 10^9 \text{ J}$ [1]

c Maximum kinetic energy at ground = loss of potential energy = $1.38 \times 10^9 \text{ J}$ [1]

Therefore maximum speed v_{max} at the ground is such that $\frac{1}{2} m v_{max}^2 = 1.38 \times 10^9 \text{ J}$ [1]
$v_{max}^2 = 2 \times 1.38 \times 10^9 \text{ J}/300 \text{ kg} = 9.20 \times 10^6 \text{ m}^2 \text{ s}^{-2}$
so $v_{max} = 3.03 \text{ km s}^{-1}$ [1]

21.3

1 $F = \dfrac{6.67 \times 10^{-11} \times 7.4 \times 10^{22} \times 2000}{(3.7 \times 10^8)^2} = 0.072 \text{ N}$ [1]

2 Least distance apart = $2.3 \times 10^{11} \text{ m} - 1.5 \times 10^{11} \text{ m} = 8.0 \times 10^{10} \text{ m}$ [1]

$F = \dfrac{6.67 \times 10^{-11} \times 6.4 \times 10^{23} \times 6.0 \times 10^{24}}{(8.0 \times 10^{10})^2} = 4.0 \times 10^{16} \text{ N}$

[1]

3 Let M_E = mass of the Earth, M_S = mass of the Sun, M_M = mass of the Moon

Force on the Earth due to the Sun = $\dfrac{GM_E M_S}{r_{ES}^2}$ where r_{ES} = the distance between the centres of the Earth and the Sun

Force on the Earth due to the Moon = $\dfrac{GM_E M_M}{r_{EM}^2}$

where r_{EM} = the distance between the centres of the Earth and the Moon

The gravitational force on the Earth due to the Sun

The gravitational force on the Earth due to the Moon

$$= \dfrac{GM_E M_S}{r_{ES}^2} \div \dfrac{GM_E M_M}{r_{EM}^2} \quad [1]$$

$$= \dfrac{GM_E M_S}{r_{ES}^2} \times \dfrac{r_{EM}^2}{GM_E M_M} = \dfrac{M_S}{r_{ES}^2} \times \dfrac{r_{EM}^2}{M_M} \quad [1]$$

$$= \dfrac{2.0 \times 10^{30}}{(1.5 \times 10^{11})^2} \times \dfrac{(3.8 \times 10^8)^2}{7.4 \times 10^{32}} = 173 = 170 \text{ (2 s.f.)} \quad [1]$$

Answers to summary questions

21.4

1 a $g = \dfrac{GM}{r^2} = \dfrac{g_s R^2}{r^2} = \dfrac{9.81 \times (6400 \times 10^3)^2}{(42\,000 \times 10^3)^2}$ [1]

       $= 0.23\,\text{N kg}^{-1}$ [1]

   b i $V = -\dfrac{GM}{10} = -\dfrac{g_s R^2}{r} = -gR = -0.23\,\text{N kg}^{-1} \times 4.2 \times 10^7\,\text{m}$ [1] $= -9.7\,\text{MJ kg}^{-1}$ [1]

      ii $E_P = mV = 200\,\text{kg} \times -9.7\,\text{MJ kg}^{-1} = -1940\,\text{MJ} = 1900\,\text{MJ (2 s.f.)}$ [1]

2 a $g = \dfrac{GM}{r^2} = -\dfrac{6.67 \times 10^{-11} \times 6.4 \times 10^{23}}{(3.4 \times 10^6)^2} = 3.7\,\text{N kg}^{-1}$ [1]

   b $v_{esc} = \sqrt{2\,g_s R} = \sqrt{2 \times 3.7 \times 3.4 \times 10^6} = 5000\,\text{m s}^{-1}$ [2]

21.5

1 a Rearranging $\dfrac{r^3}{T^2} = \dfrac{GM}{4\pi^2}$ gives $r^3 = \dfrac{GMT^2}{4\pi^2}$ where $T = 90 \times 60\,\text{s} = 5400\,\text{s}$

      $r^3 = \dfrac{GMT^2}{4\pi^2} = \dfrac{6.670 \times 10^{-11} \times 6.0 \times 10^{24} \times (5400)^2}{4\pi^2}$ [1]

      $= 2.96 \times 10^{20}\,\text{m}^3$

      so $r = 6.7 \times 10^6\,\text{m}$ [1]

      Height above the ground = $6.7 \times 10^6\,\text{m} - 6.4 \times 10^6\,\text{m} = 300\,\text{km}$ [1]

   b Centripetal acceleration $= v^2/r = GM/r^2$ [1] so

      $v = \sqrt{\dfrac{GM}{r}} = \sqrt{\dfrac{6.67 \times 10^{-11} \times 6.0 \times 10^{24}}{6.7 \times 10^6}}$ [1]

      $= 7.7\,\text{km s}^{-1}$ [1]

2 a Rearranging $\dfrac{r^3}{T^2} = \dfrac{GM}{4\pi^2}$ gives $T^2 = \dfrac{4\pi^2 r^3}{GM}$ where $r = 7600\,\text{km}$

      $T^2 = \dfrac{4\pi^2 r^3}{GM} = \dfrac{4\pi^2 \times (7.6 \times 10^6)^3}{6.67 \times 10^{-11} \times 6.0 \times 10^{24}}$ [1]

      $= 4.33 \times 10^7\,\text{s}^2$ so $T = 6600\,\text{s}$ [1]

   b Angular speed $= 2\pi/T = 9.5 \times 10^{-4}\,\text{rad s}^{-1}$ [1]

22.1/22.2

1 a The field lines are directed from the plate to Q [1]

   b i Vertically down [1]

      ii Downwards at about 45° to the vertical. [1]

2 a No of electrons $n = Q/e = 2.5\,\text{nc}/1.6 \times 10^{-19}\,\text{C} = 1.56 \times 10^{10}$ [1]

      Surface area $A = 4\pi R^2 = 4\pi \times (0.15\,\text{m})^2 = 0.283\,\text{m}^2 = 2.83 \times 10^5\,\text{mm}^2$ [1]

      No of electrons per $\text{mm}^2 = n/A = 1.56 \times 10^{10}/2.83 \times 10^5\,\text{mm}^2 = 5.5 \times 10^4\,\text{mm}^{-2}$ [1]

   b $t = Q/I = 2.5\,\text{nC}/0.2\,\text{mA} = 12.5\,\mu\text{s} = 13\,\mu\text{s (2 s.f.)}$ [1]

3 a $F = QE = 3 \times 1.6 \times 10^{-19}\,\text{C} \times 900\,000\,\text{V m}^{-1}$ [1]

      $= 4.3 \times 10^{-13}\,\text{C}$ [1]

   b $V = E\,d = 900\,000\,\text{V m}^{-1} \times 0.050\,\text{m}$ [1] $= 45\,\text{kV}$ [1]

22.3

1 a $W = F\,\Delta x = QE \times \Delta x = 1.6 \times 10^{-19}\,\text{C} \times 8000\,\text{V m}^{-1} \times 0.028\,\text{m}$ [1] $= 3.6 \times 10^{-17}\,\text{J}$ [1]

   b $\Delta V = E\,\Delta x = 8000\,\text{V m}^{-1} \times 0.028\,\text{m} = 224\,\text{V}$ [1]
      Potential at Y = $+400\,\text{V} + 224\,\text{V} = +624\,\text{V}$ [1]

2 a $W = Q\Delta V = +2e\,(6000\,\text{V} - 1000\,\text{V})$ [1]
      $= 2 \times 1.6 \times 10^{-19}\,\text{C} \times 5000\,\text{V} = 1.6 \times 10^{-15}\,\text{J}$ [1]

   b Average potential gradient $= \Delta V/\Delta x = 5000\,\text{V}/0.036\,\text{m}$ [1] $= 1.4 \times 10^5\,\text{V m}^{-1}$ [1]

3 a Graph should be a straight line with a negative gradient [1] from $y = +4200\,\text{V}$ at $x = 0$ at L to $y = 0\,\text{V}$ at $x = 80\,\text{mm}$ at m [1]

   b i Potential gradient $= \Delta V/\Delta x = (4200 - 0)\,\text{V}/(0 - 0.080\,\text{m}) = -5.3 \times 10^4\,\text{V m}^{-1}$ [1]

      ii Electric field strength = $-$potential gradient $= +5.3 \times 10^4\,\text{V m}^{-1}$ [1] from L to M [1]

22.4/22.5

1 $F = \dfrac{1}{4\pi\varepsilon_o}\dfrac{Q_1 Q_2}{r^2}$

   $= \dfrac{1}{4\pi(8.85 \times 10^{-12}\,\text{F m}^{-1})} \times \dfrac{(2 \times 1.6 \times 10^{-19}\,\text{C})^2}{(5.6 \times 10^{-7}\,\text{m})^2}$ [1]

   $= 2.9 \times 10^{-15}\,\text{m}$ [1]

2 The initial force $F_1 = \dfrac{1}{4\pi\varepsilon_o}\dfrac{Q_X Q_Y}{r^2}$ where $r = 1.6 \times 10^{-2}\,\text{m}$

   The force increases to $F_2 = 7.8 \times 10^{-5}\,\text{N}$ from $5.1 \times 10^{-5}\,\text{N}$, which is an increase of $\times 1.53$ ($= 7.8 \times 10^{-5}\,\text{N}/5.1 \times 10^{-5}\,\text{N}$) [1]. So the charge on X must have increased by $\times 1.53$ [1]

   $\dfrac{(Q_X + q)}{Q_X} = 1.53$ where q is the increase of charge so

   $(Q_X + q) = 1.53 Q_X$ therefore $Q_X = q/0.53 = 2.1\,\text{nC}/0.53 = 4.0\,\text{nC}$ [1]

   Rearrange $F_1 = \dfrac{1}{4\pi\varepsilon_o}\dfrac{Q_X Q_Y}{r^2}$ to give

   $Q_Y = \dfrac{4\pi\varepsilon_o r^2 F_1}{Q_X}$

   $= \dfrac{4\pi \times 8.85 \times 10^{-12}\,\text{F m}^{-1} \times (1.6 \times 10^{-2}\,\text{m})^2 \times (5.1 \times 10^{-5}\,\text{N})}{4.0 \times 10^{-9}\,\text{C}}$

   [1]

   $= 3.6 \times 10^{-10}\,\text{C} = 0.36\,\text{nC}$ [1]

3 a Potential energy $= \dfrac{1}{4\pi\varepsilon_o}\dfrac{Qq}{r}$

      $= \dfrac{1}{4\pi \times 8.85 \times 10^{-12}\,\text{F m}^{-1}} \times \dfrac{(+1.6 \times 10^{-19}\,\text{C})(+1.6 \times 10^{-19}\,\text{C})}{0.11 \times 10^{-9}}$ [1] $= 2.1 \times 10^{-18}\,\text{J}$ [1]

Answers to summary questions

b The distance from the midpoint to each charge = 0.055 nm

Electric field strength at the midpoint due to the proton $E_1 = \dfrac{1}{4\pi\varepsilon_o}\dfrac{+e}{r^2}$

$= \dfrac{+1.6\times 10^{-16}\,\text{C}}{4\pi\times 8.85\times 10^{-12}\,\text{F m}^{-1}\times(0.055\times 10^{-9}\,\text{m})^2}$

$= 4.8 \times 10^{11}\,\text{V m}^{-1}$ away from the proton [1]

Electric field strength at the midpoint due to the electron $E_2 = \dfrac{1}{4\pi\varepsilon_o}\dfrac{+e}{r^2} = 4.8\times 10^{11}\,\text{V m}^{-1}$ towards the electron [1]

Resultant electric field strength = $2 \times 4.8 \times 10^{11}\,\text{V m}^{-1} = 9.6 \times 10^{11}\,\text{V m}^{-1}$ towards the electron [1]

4 a i Rearranging $V = \dfrac{1}{4\pi\varepsilon_o}\dfrac{Q}{r}$ gives $Q = 4\pi\varepsilon_o rV =$

$4\pi\times 8.85\times 10^{-12}\,\text{F m}^{-1}\times 80\,\text{mm}\times(+6200\,\text{V})$ [1]

$= +55\,\text{nC}$ [1]

ii $V \propto 1/r$ and r is increased by $\times 12.5$ ($= 1000\,\text{mm}/80\,\text{mm}$) [1]

Therefore at 1000 mm, $V = +6.2\,\text{kV}/12.5$ ($= 496\,\text{V}$) = 500 V to 2 significant figures. [1]

b $E = Q/4\pi\varepsilon_o r^2$ and $V = Q/4\pi\varepsilon_o r$ Therefore $E = V/r = 6200\,\text{V}/0.080\,\text{m}$ [1] $= 77500\,\text{V m}^{-1}$ $= 7.8\,\text{kV m}^{-1}$ to 2 significant figures. [1]

22.6

1 Any 2 of the following:
- field lines are parallel in a uniform field but not in a radial field
- field lines of an object's gravitational field are always towards the object whereas the direction of the field lines in an electric field depends on whether the charge is + or –
- the field strength in a uniform field is the same at all points in the field which is not so for a radial field
- the equipotentials in a radial field are spherical (or circular) whereas in a uniform field the equipotentials are flat surfaces (or straight lines) [2]

2 Both field strength rows correct [1] Both potential rows correct [1]

	Z	Y	k	n
Gravitational field strength	g	M	G	-2
Electric field strength	E	Q	$\dfrac{1}{4\pi\varepsilon_o}$	-2
Gravitational potential	V	M	G	-1
Electric potential	V	Q	$\dfrac{1}{4\pi\varepsilon_o}$	-1

3 a Rearranging $F = \dfrac{Q}{4\pi\varepsilon_o r^2}$ gives $\dfrac{1}{4\pi\varepsilon_o} = F r^2 Q^{-1}$ [1]

$= \text{kg m s}^{-2}\,\text{m}^2\,\text{A}^{-1}\,\text{s}^{-1} = \text{kg m}^3\,\text{s}^{-3}\,\text{A}^{-1}$ [1]

b $\sqrt{(2ER)}$ [1]

23.1/23.2/23.3

1 a $Q = CV = 47\,\mu\text{F} \times 6.0\,\text{V} = 2.8\times 10^{-4}\,\text{C}$ [1]

b $Q = It$ gives $t = \dfrac{Q}{I} = \dfrac{2.8\times 10^{-4}\,\text{C}}{1.5\times 10^{-6}\,\text{C}} = (187\,\text{s} =)$ 190 s to 2 significant figures [1]

2 a $Q = CV = 22\,\mu\text{F}\times 3.0\,\text{V} = 66\,\mu\text{C}$ [1]

b i Charge transferred from the 22 µF capacitor to the 10 µF capacitor [1]. (As the total charge is conserved), the 10 µF capacitor gains charge and the 22 µF capacitor loses an equal amount of charge. [1]

ii The pd across the 10 µF capacitor = $q/10\,\mu\text{F}$ where q is the charge transferred [1]

The pd across the 22 µF capacitor = $(66\,\mu\text{C} - q)/22\,\mu\text{F}$ [1]

The pd across the 2 capacitors is the same therefore $q/10\,\mu\text{F} = (66\,\mu\text{C} - q)/22\,\mu\text{F}$ [1]

$2.2q = 66\,\mu\text{C} - q$ which gives $3.2q = 66$ so $q = 20.6\,\mu\text{C}$

Therefore the pd across the 10 µF capacitor = $20.6\,\mu\text{C}/10\,\mu\text{F} = 2.1\,\text{V}$ to 2 significant figures [1]

3 a i $Q = CV = 10\,\mu\text{F}\times 6.0\,\text{V} = 60\,\mu\text{C}$ [1]
$E = \tfrac{1}{2}CV^2 = 0.5\times 10\,\mu\text{F}\times(6.0\,\text{V})^2$ (or $\tfrac{1}{2}QV$) $= 180\,\mu\text{J}$ [1]

ii Energy supplied by the battery $= QV = 60\,\mu\text{C}\times 6.0\,\text{V} = 360\,\mu\text{J}$ [1]

b The difference is because energy is dissipated due to resistance heating when charge flows round the circuit. [1] The energy supplied by the battery is the sum of the energy stored in the capacitor and the energy dissipated due to resistance heating in the circuit. [1]

4 a i Time constant $= RC = 200\,\text{k}\Omega\times 47\,\mu\text{F} = 9.4\,\text{s}$ [1]

ii The time constant is the time taken to discharge to 37% of the initial pd which is 2.2 V in this case. [1] So the capacitor takes about 10 s (just over 9.4 s) for the pd to decrease to 2 V [1]

b $V = V_o e^{-\tfrac{t}{RC}} = 6.0\,\text{V}\times e^{-\tfrac{5.0\,\text{s}}{9.4\,\text{s}}}$ [1]

$= 6.0\,\text{V}\times 0.59 = 3.5\,\text{V}$ [1]

Current $= V/R = 3.5\,\text{V}/200\,\text{k}\Omega = 18\,\mu\text{A}$ [1]

5 a i Current $= V/R = 5.0\,\text{V}/20\,\text{k}\Omega = 2.5\times 10^{-4}\,\text{A}$ [1]

ii The pd across the resistor $= 5.0\,\text{V} - 3.0\,\text{V} = 2.0\,\text{V}$ [1] so the current = resistor pd/resistance $= 2.0\,\text{V}/20\,\text{k}\Omega = 1.0\times 10^{-4}\,\text{A}$ [1]

b The time for the pd to increase to 3.0 V = time for the current to decrease to $1.0\times 10^{-4}\,\text{A}$ from $2.5\times 10^{-4}\,\text{A}$ [1]

Rearranging $I = I_o e^{-\tfrac{t}{RC}}$ gives $\ln(I/I_o) = -t/RC$ [1]

Therefore $t = -RC\ln(I/I_o) = -20\,\text{k}\Omega\times 68\,\mu\text{F}\times \ln\dfrac{1.0\times 10^{-4}\,\text{A}}{2.5\times 10^{-4}\,\text{A}} = 1.2\,\text{s}$ to 2 significant figures [1]

Answers to summary questions

23.4

1 a i The capacitance decreases [1] because it is inversely proportional to the distance between the plates. [1]

   ii The pd between the plates increases. [1] This is because the charge Q is unchanged (as the capacitor plates are isolated) and the capacitance C decreases so $V = Q/C$ increases. [1]

   b Work is done to overcome the force of attraction between the plates when the plates are moved apart [1]. So the energy stored increases by an amount equal to the work done. [1]

2 The molecules are polarised by the electric field to become dipoles (if they are not already polarised). [1] The dipoles rotate to align with the field as their positive pole is attracted to the negative plate and their negative pole is attracted to the positive plate. [1] The effect is to create a layer of positive charge against the negative plate and a layer of negative charge against the positive plate. [1] This 'polarisation' charge allows more electrons to move onto the negative plate from the battery and off the positive plate back to the battery. [1]

24.1

1 a $B = \dfrac{F}{Il} = \dfrac{0.028\,\text{N}}{4.3\,\text{A} \times 0.080\,\text{m}}$ [1] $= 0.081\,\text{N}$ [1]

   b Eastwards and horizontal [1]

2 Arrangement as Figure 1 [1]
magnetic field IN

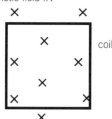

▲ Figure 1

Magnitude of the force on each side $= BIln = 0.055\,\text{T} \times 5.9\,\text{A} \times 0.060\,\text{m} \times 50$ [1] $= 0.97\,\text{N}$ [1]

Direction of the force on each side is perpendicular to the side and horizontal [1]

3 Force per unit length $= BI = 6.6 \times 10^{-5}\,\text{T} \times 65\,\text{A}$ [1]
$= 4.3 \times 10^{-3}\,\text{N m}^{-1}$ [1]

24.2

1 a $F = Bev = 0.150\,\text{T} \times 1.6 \times 10^{-19}\,\text{C} \times 8.0 \times 10^{6}\,\text{s}^{-1}$
$= 1.9 \times 10^{-13}\,\text{N}$ [1]

   b The force is zero because the direction of motion of the electrons is parallel to the magnetic field. [1]

2 a $Bqv = mv^2/r$ hence $r = \dfrac{mv}{Bq}$ [1]

$= \dfrac{1.67 \times 10^{-27}\,\text{kg} \times 3.2 \times 10^{7}\,\text{m s}^{-1}}{0.085\,\text{T} \times 1.6 \times 10^{-19}\,\text{C}}$ [1] $= 3.9\,\text{m}$ [1]

   b From $r = \dfrac{mv}{Bq}$, the radius of curvature is proportional to $\dfrac{v}{B}$ [1]

The ratio $\dfrac{\text{the new magnetic flux density } B_2}{\text{the initial magnetic flux density } B_1} = 0.5$

The ratio $\dfrac{\text{the new speed } v_2}{\text{the initial speed } v_1} = \sqrt{\dfrac{\text{the new KE}}{\text{the initial KE}}}$
$= 0.71$ [1]

Therefore $\dfrac{\text{the new radius of curvature } r_2}{\text{the new radius of curvature } r_1} =$

$\dfrac{v_2}{B_2} \div \dfrac{v_1}{B_1} = \dfrac{v_2}{v_1} \sqrt{\dfrac{B_2}{B_1}} = 0.71 \div 0.5 = 1.4$

The new radius of curvature is 1.4 × the initial radius of curvature [1]

3 a $Bqv = qE$ gives $B = \dfrac{E}{v} = \dfrac{8.2 \times 10^{4}\,\text{V}}{9.5 \times 10^{5}\,\text{m s}^{-1}}$ [1]
$= 0.086\,\text{T}$ [1]

   b From $r = \dfrac{mv}{Bq}$,

specific charge $\dfrac{q}{m} = \dfrac{v}{Br}$ [1]

$= \dfrac{9.5 \times 10^{5}\,\text{m s}^{-1}}{0.91\,\text{T} \times 0.185\,\text{m}} =$ [1]

$= 5.6 \times 10^{6}\,\text{C kg}^{-1}$ [1]

25.1

1 a An emf is induced in the rod because it cuts across the magnetic field lines. [1] The rod is in a complete circuit with the meter so there is an induced current in the circuit while the rod is moving [1]

   b Either move the rod faster (or use a stronger magnet) [1]

2 The induced current is upwards from Y to X [1]. See Figure 1 below for the application of the right-hand rule in this situation.

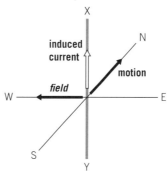

▲ Figure 1

Answers to summary questions

Conduction electrons in the rod move from south to north due to the movement of the rod. The magnetic field exerts a force on the moving electrons and forces them towards Y away from X [1]

3 The magnetic flux linkage in the coil decreases to zero and this induces an emf in the coil which causes a current in the circuit. [1] Once the coil has been removed from the field (and stays outside the field) no further change of flux occurs so the meter reading is zero. [1]

When the coil is returned slowly to the magnetic field, the flux linkage through it increases so an emf is induced with opposite polarity to when it was removed so the meter deflects in the opposite direction. [1] The change takes longer so the induced current is less than when the coil was removed. [1]

25.2

1 a Area swept out each second by the rod = 0.063 m × 0.012 m = 7.56 × 10⁻⁴ m² [1]

Magnetic flux swept out each second = $B \times A/t$ = 0.072 T × 7.56 × 10⁻⁴ m² [1] = 5.4 × 10⁻⁵ Wb [1]

b Induced emf = flux change per second = 5.4 × 10⁻⁵ Wb ÷ 1.0 s = 5.4 × 10⁻⁵ V [1]

2 a The magnetic flux linkage in the solenoid due to the bar magnet increases as the magnet approaches the solenoid [1] so an emf is induced in the solenoid [1]. The solenoid and the meter form a complete circuit so the induced emf causes an induced current in the circuit. [1]

b The meter deflects in the opposite direction as the magnet is being moved away. This is because the magnetic flux linkage in the solenoid due to the bar magnet decreases as the magnet approaches the solenoid [1] so an emf is induced in the solenoid and this causes an induced current in the circuit. [1]. The current is in the opposite direction because the induced current creates a magnetic field which opposes the magnet moving away from the coil [1]

25.3/4

1 a Maximum flux linkage = $B A n$
= 0.110 T × 0.022 m × 0.018 m × 120 [1]
= 5.2(3) × 10⁻³ Wb [1]

b Maximum emf = $B A n \omega$ = 5.2(3) × 10⁻³ Wb × 2π × 50 Hz [1] = 1.6(4) V [1]

2 When the motor is spinning, an emf is induced in the coil that opposes its motion. The faster the motor spins, the greater this back emf is. [1] When the load is increased, the coil's frequency of rotation is reduced so the back emf is reduced. [1]

The current in the coil = (the supply pd − the back emf) ÷ coil resistance so the current increases because the back emf decreases. [1]

3 a rms current = power ÷ pd = 1000 W ÷ 230 V = 4.3(5) A [1]

b The peak current = $\sqrt{2}$ × 4.35 V = 6.1 A [1]

c The peak power = peak current × peak pd = 6.1 A × (230√2) V [1] = 2000 W [1]

25.5

1 When the device is switched on, the magnetic flux in the core is due to the difference between the flux due to the primary coil and the flux due to the secondary coil. [1]

Either

When the device is switched off, the current in the secondary coil becomes zero and no power is transferred to the secondary coil from the primary coil. [1] The primary current decreases because less power is supplied to the primary coil. [1]

or

The magnetic flux in the core increases as it is now due only to the primary coil as there is no longer any oppositely-directed flux due to the secondary coil. [1] So the back emf in the primary coil increases and reduces the primary current. [1]

2 a The current in the coils has a heating effect due to their resistance. [1]

Eddy currents induced in the core have a heating effect due to the resistance of the core [1]

Energy is wasted each time the core is magnetised and demagnetised. [1]

b Any 4 out of the following 5 points;-

Power = current × pd. [1] By increasing the grid voltage, less current is needed to transfer the same amount of power. [1] With less current, less power is wasted due to resistance heating by the current in the grid cables. [1] So less power needs to be supplied to the grid cables to deliver the same amount of power to the load connected to the cables. [1] The system efficiency is equal to the power delivered to the load ÷ the power supplied to the cables. [1]

3 a $N_S = N_P \times \dfrac{V_S}{V_P} = 2400 \times \dfrac{6.0\,\text{V}}{230\,\text{V}}$ [1] = 63 turns [1]

b Assume the efficiency = 100% so $I_S V_S = I_P V_P$ [1]
$I_P = \dfrac{I_S V_S}{V_P} = \dfrac{24\,\text{W}}{230\,\text{V}}$ [1] = 0.10(4) A [1]

26.1

1 a The atom is mostly empty space [1] and most of its mass is concentrated in a nucleus which is much smaller than the size of the atom [1]. The nucleus is positively charged because it repels and deflects α particles which are positively charged [1]

b Any two points

Particles moving at different speeds moving along the same initial path would be deflected by different amounts. [1]

The detector reading at any given angle of deflection (other than 180°) would be due to particles moving along different paths at different speeds. [1]

The nuclear model of the atom could not have been proved from the measurements. [1]

2 a See Figure 1

Initial and final directions at 90°

with N in top RH quadrant [1]

P in LH bottom quadrant at midpoint of curve

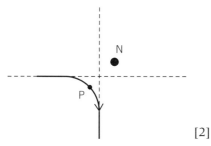

[2]

▲ Figure 1

b Any 3 points

E_K decreases to a minimum at P then increases after P [1]

E_P increases to a maximum at P then decreases after P [1]

Initial E_K = Final E_K [1]

At any point $E_K + E_P$ = initial E_K [1]

3 $E_K = \dfrac{Q_\alpha Q_N}{4\pi\varepsilon_0 d}$ gives $d = \dfrac{Q_\alpha Q_N}{4\pi\varepsilon_0 E_K}$

$= \dfrac{2e \times 13e}{4\pi \times 8.85 \times 10^{-12}\,F\,m^{-1} \times 5.1 \times 10^6\,e}$ [1]

$= \dfrac{26 \times 1.6 \times 10^{-19}\,C}{4\pi \times 8.85 \times 10^{-12}\,F\,m^{-1} \times 5.1 \times 10^6}$ [1]

$= 7.3 \times 10^{-15}\,m$ [1]

26.2/3

1 a A proton in the nucleus changes into a neutron. [1]

The number of protons in the nucleus decreases by 1 and the number of neutrons increases by 1. [1]

b The count rate C is proportional to the intensity I. From the inverse-square law equation I is proportional to $1/r^2$, so $C = k/r^2$

$k = C r^2 = 25.2\,s^{-1} \times (0.075\,m)^2$ [1]

At 160 mm, $C = 25.2\,s^{-1} \times (0.075\,m)^2/(0.160\,m)^2$ [1] = 5.5 counts per second [1]

2 a $^{210}_{85}At + ^{\ 0}_{-1}e$ [1] $\to ^{210}_{84}Po + v_e$ [1]

b $^{210}_{84}Po \to ^{4}_{2}\alpha + ^{206}_{82}Pb$ [1]

26.4

1 α radiation is highly-ionising radiation [1] and when it passes through the body from a source inside the body it creates ions in the body [1] that can kill or damage body cells.

α radiation reaching the body from a source outside the body is stopped by the outer layers of the skin so is much less damaging than if the source is in the body. [1]

2 a Provided the lead lining is thick enough [1], it absorbs all the radiation from the source (including γ radiation) inside the box so no radiation penetrates to the outside of the box. [1]

b The dose of ionising radiation received by the user depends on how long the user is exposed to the radiation. [1] By making the exposure time as short as possible, the risk to the user is reduced. [1]

3 Any 2 points

Ionisation in a living cell can kill cells by destroying cell membranes [1] or damage DNA molecules in cells causing them to divide and grow and create tumours [1] or cause a mutation in a sex cell which may be passed on to future generations. [1]

26.5/6

1 a 8 hours = 4 half-lives so the number of atoms = $(0.5)^4$ [1] × the initial number = 1.1×10^{14} [1]

b 3 hours = 1.5 half-lives so the number of atoms = $(0.5)^{1.5}$ [1] of the initial number = 6.4×10^{14} [1]

2 a The decay constant of an isotope is the probability per unit time of a nucleus (of the isotope) decaying [1]

b i decay constant $\lambda = \ln 2/T_{1/2}$
$= \ln 2/(1620 \times 365 \times 24 \times 3600\,s)$
$= 1.36 \times 10^{-11}\,s^{-1}$ [1]
number of atoms per kg = $6.02 \times 10^{23}/0.226\,kg$
$= 2.67 \times 10^{24}\,kg^{-1}$ [1]
initial activity = $\lambda N = 1.36 \times 10^{-11}\,s^{-1} \times 2.67 \times 10^{24} = 3.6 \times 10^{13}$ Bq [1]

ii $m = m_0 e^{-\lambda t}$ where $\lambda t = 1.36 \times 10^{-11}\,s^{-1} \times 3000 \times 365 \times 24 \times 3600\,s = 1.29$ [1]

Therefore $m = 1.0\,kg \times e^{-1.29} = 0.28\,kg$ [1]

OR 3000 yrs = 3000/1620 half-lives = 1.85 half-lives [1] therefore $m = 1.0\,kg \times 0.5^{1.85} = 0.28\,kg$. [1]

iii Number of atoms remaining = $2.67 \times 10^{24}\,kg^{-1} \times 0.28\,kg = 7.5 \times 10^{23}$ [1]

3 a Rearranging $A = A_0 e^{-\lambda t}$ gives $\lambda t = \ln(A_0/A)$ [1]
$= \ln(40\,kBq/36\,kBq) = \ln 1.11 = 0.105$ [1]
$\lambda = 0.105/24\,h = 0.105/(24 \times 3600\,s) = 1.2(2) \times 10^{-6}\,s^{-1}$ [1]

b $T_{1/2} = \ln 2/\lambda = 0.693/1.2(2) \times 10^{-6}\,s^{-1}$ [1]
$= 5.7 \times 10^5\,s$ [1]

c $A = A_0 e^{-\lambda t}$ where $\lambda t = 1.2(2) \times 10^{-6}\,s^{-1} \times 120 \times 3600\,s$
$= 0.527$ [1]

Therefore $A = 40\,kBq \times e^{-0.527}$ [1] = 24 kBq [1]

26.7

1 a For each U-238 nucleus now present, 6.4 U-238 nuclei decayed and formed Pb-206 nuclei. So the proportion of U-238 now to when the rock formed is 1 : 7.4 [1]

Answers to summary questions

b $\lambda = \ln 2/(4500 \times 10^6 \times 365 \times 24 \times 3600\,s) = 4.88 \times 10^{-18}\,s^{-1}$ [1]

Rearranging $N = N_0 e^{-\lambda t}$ gives $t = \ln(N_0/N) \div \lambda$ [1]
$= [\ln(7.4/1)] \div 4.88 \times 10^{-18}\,s^{-1}$ [1]
$= 4.10 \times 10^{17}\,s = 1.3 \times 10^{10}$ years [1]

2 a Number of Pu-238 nuclei in 0.20 g = $(6.02 \times 10^{23}/238\,g) \times 0.20\,g = 5.1 \times 10^{20}$ [1]

$\lambda = \ln 2/(88 \times 365 \times 24 \times 3600\,s)$
$= 2.50 \times 10^{-10}\,s^{-1}$ [1]

Activity $= \lambda N = 2.50 \times 10^{-10}\,s^{-1} \times 5.1 \times 10^{20}$
$(= 1.28 \times 10^{11}\,Bq) = 1.3 \times 10^{11}\,Bq$ [1]

b Energy release per second $= AE = 1.28 \times 10^{11}\,Bq \times 5.6 \times 1.6 \times 10^{-13}\,J$ [1] $= 0.11\,J\,s^{-1}$ [1]

26.8

1 a See Topic 26.8 Figure 1 straight line from origin to approx $Z = A = 20$ [1]; line curves towards N axis and finishes about $Z = 80$, $N = 120$ [1]

b i near $Z = 80$, $N = 120$ and above [1]

ii near $N = Z$ curve on N-axis side [1]

iii near $N = Z$ curve on Z-axis side [1]

2 a Emission of 8 α particles (reduces A by 32 to 206 and) reduces Z by 16 to 76. [1] Each $β^-$ emission increases Z **by 1** so there must be 6 $β^-$ emissions to increase Z from 76 to 82. [1]

b Each α emission moves the nuclide towards the N-axis and away from the N-Z curve of stable nuclei [1]. $β^+$ emission takes place in nuclei that are on the other side of the $N=Z$ plot to where the nuclide has moved to. [1] α emission increases the neutron to proton ratio which means $β^-$ emission not $β^+$ emission could then take place to move back towards stability [1]

3 a See Figure 2

[1] for top and bottom energy level values shown

[1] for 1.35 MeV level with energy shown

[1] for 0.47 MeV level with energy shown

[1] for all 4 correct transitions shown

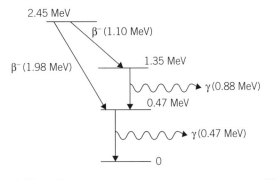

▲ Figure 2 [4]

b $^{24}_{10}Ne \rightarrow\,^{0}_{-1}β + ^{24}_{11}Na$ [1] $+ \bar{v}_e$ [1]

26.9

1 $\lambda = hc/E = (6.63 \times 10^{-34}\,J\,s \times 3.0 \times 10^8\,m\,s^{-1})/(310 \times 1.6 \times 10^{-13}\,J) = 4.0 \times 10^{-15}\,m$ [1]
$R \sin \theta_{min} = 0.61\lambda$ gives $R = 0.61\lambda/\sin \theta_{min}$
$= 0.61 \times 4.0 \times 10^{-15}\,m/\sin 39° $ [1] $= 3.9 \times 10^{-15}\,m$ [1]

2 The angle of diffraction was between 37° and 41°. Since $\sin 37° = 0.6018$, $\sin 39° = 0.6293$, and $\sin 40° = 0.6561$, the % uncertainty in $\sin \theta = (0.6561 - 0.6018/2) \div 0.6293 \times 100\% = 4.3\%$. [1] Therefore % uncertainty in $R = 4.3\%$ so uncertainty in $R = \pm 0.043 \times 3.9\,fm = \pm 0.2\,fm$. [1]

3 $R = r_0 A^{1/3} = 1.05\,fm \times 28^{1/3} = 3.2\,fm$ [1]
volume $= 4\pi R^3/3 = 4\pi (3.2 \times 10^{-15}\,m)^3/3 = 1.37 \times 10^{-43}\,m^3$

mass $= 28\,u = 28 \times 1.661 \times 10^{-27}\,kg = 4.65 \times 10^{-26}\,kg$ [1]

density $=$ mass/volume $= 4.65 \times 10^{-26}\,kg \div 1.37 \times 10^{-43}\,m^3$
$= 3.4 \times 10^{17}\,kg\,m^{-3}$ [1]

27.1

1 a $^{14}_{6}C \rightarrow\,^{14}_{7}N$ [1] $+\,^{0}_{-1}β$ [1] $+ \bar{v} + Q$

b mass defect $\Delta m = 13.99995\,u - 13.99923\,u - 0.00055\,u = 0.00017\,u$ [1]

Energy released $Q = 931.5\,MeV/u \times 0.00017\,u$ [1] $= 0.16\,MeV$ [1]

2 a $^{37}_{18}Ar +\,^{0}_{-1}e \rightarrow\,^{37}_{18}Cl$ [1] $+ v$ [1] $+ Q$ [2]

b mass defect $\Delta m = 36.95689\,u + 0.00055\,u - 36.95657 = 0.00087\,u$ [1]

Energy released $Q = 931.5\,MeV/u \times 0.00087\,u$ [1] $= 0.81\,MeV$ [1]

3 a $^{241}_{95}Am \rightarrow\,^{237}_{93}Np$ [1] $+\,^{4}_{2}α$ [1] $+ Q$

b mass of $^{241}_{95}Am = 241.05669 - (95 \times 0.00055)$
$= 241.00444\,u$ [1]

mass of $^{237}_{93}Np = 237.04803 - (93 \times 0.00055) = 236.99688\,u$ [1]
mass defect $\Delta m = 241.00444\,u - 236.99688\,u - 4.00150\,u = 0.00606\,u$ [1]

Energy released $Q = 931.5\,MeV/u \times 0.00606\,u$ [1] $= 5.6\,MeV$ [1]

27.2

1 a The binding energy of a nucleus is the work done to separate its protons and neutrons from each other completely. [1]

The greater the binding energy of a nucleus is, the more stable it is. [1]

b The binding energy of the nucleus increases because it releases energy (or it is less unstable or more tightly bound after the emission) [1]. So the binding energy per nucleon of the nucleus increases. [1]

Answers to summary questions

2 a Mass defect $\Delta m = 2m_p + m_n - M_{NUC} = 0.00726$ u [1]

   Therefore, binding energy = 0.00726 u × 931.5 MeV/u = 6.76 MeV [1]

   Binding energy per nucleon = 6.76 MeV/3 = 2.3 MeV [1]

   b Mass defect $\Delta m = 92m_p + 146m_n - M_{NUC} = 1.885$ u [1]

   Therefore, binding energy = 1.885 u × 931.5 MeV/u = 1756 MeV [1]

   Binding energy per nucleon = 1756 MeV/238 = 7.4 MeV [1]

3 a $^2_1H + ^1_1p$ [1] $\rightarrow ^3_2He$ [1] + Q

   b Mass difference Δm = 2.01355 u + 1.00728 u − 3.01493 u = 0.00590 u [1]

   Energy released Q = 931.5 MeV/u × 0.00590 u [1] = 5.5 MeV [1]

27.3

1 a The strong nuclear force between the protons and neutrons in the nucleus holds them together [1] because it is an attractive force and prevents the coulomb repulsion force between the protons pushing the protons out of the nucleus. [1]

   b The work done by the strong nuclear force when the nucleus is formed increases the binding energy of the protons and neutrons when they come together. [1] The mass of the protons, and neutrons is therefore reduced because some of their mass is converted to binding energy [1] in accordance with the equation $E = mc^2$ [1]

2 Mass before = 234.94342 + 1.00867 = 235.95209 u [1]

   Mass after = 139.89192 + 92.89313 + (3 × 1.00867) = 235.81106 u [1]

   Mass difference = 0.141 u [1]

   Energy released = 931.5 MeV/u × 0.141 u = 131 MeV [1]

3 a $^3_1H + ^2_1H$ [1] $\rightarrow ^4_2He + ^1_0n$ [1] + Q

   b Mass before = 3.01550 + 2.01355 = 5.02905 u [1]

   Mass after = 4.00150 + 1.00867 = 5.01017 u [1]

   Mass difference = 0.01888 u [1]

   Energy released = 931.5 MeV/u × 0.01888 u = 17.6 MeV [1]

27.4

1 a The moderator reduces the average speed (or kinetic energy) of the fission neutrons [1] so they can produce further fission. [1]

   b The fission neutrons repeatedly collide with the moderator atoms. [1] The collisions are elastic [1] and in each collision some of the kinetic energy of the fission neutron is transferred to the moderator atom. [1] The percentage of the kinetic energy transferred to the moderator atom from the fission neutron depends on the mass of the moderator atom. [1] The closer the atomic mass of the moderator is to the mass of the neutron, the greater the percentage of the kinetic energy transferred to the moderator atom. [1]

2 a The control rods absorb neutrons. [1] For energy to be released at a constant rate, the depth of the control rods in the reactor core is adjusted (automatically) [1] so that an average of one fission neutron per fission event causes a further fission event. [1]

   b The control rods are pushed/inserted fully into the reactor core [1] so the rate at which they absorb neutrons in increased. [1] As a result, the number of fission neutrons in the reactor core decreases [1] and the decrease continues until fission stops. [1]

3 The fuel rods consist of the uranium isotopes $^{235}_{92}U$ (which is fissile) and $^{238}_{92}U$ [1]. Both isotopes emit α particles and these particles are unable to pass through the metal walls of the fuel. [1]

   The used fuel rods contain many different neutron-rich fission fragments which emit β⁻ radiation and γ radiation, [1] both of which can pass through the metal walls [1]. In addition, the used fuel rods contain highly-radioactive plutonium $^{239}_{94}Pu$ formed as a result of $^{238}_{92}U$ nuclei absorbing neutrons then decaying by β⁻ emission. [1] So more radiation passes through the metal walls of the fuel rods after they have been used than before

Astrophysics

1.1

1 a $v = -0.240$ m, image height = 54 mm [2]

   b virtual, upright, enlarged [1]

1.2/1.5

1 a i 8.0 [1]   ii 1.20° [1]

   b A parabolic mirror focuses all rays parallel to the axis to a single point whereas a convex lens does not [1] A mirror reflects light whereas a lenses refracts it so chromatic aberration cannot occur with a mirror. [1]

   c 6.3×10^{-5} rad [1] = 3.6×10^{-3} degrees [1]

2.1/2.2

1 a The absolute magnitude, M, of a star is its apparent magnitude, m, if it was at a distance of 10 parsecs from Earth. [1]

   b $m - M = 5 \log(d/10)$ gives $M = +4.8 - 5 \log(34/10)$ [1] = +2.1 [1]

Answers to summary questions

2.3

1   a   P has the greater output power. [1] It is a red giant and will become a white dwarf before it becomes invisible. [1] Q is a main sequence star [1] that will become a red giant then a white dwarf. [1]

   b   Their magnitude difference is 8 so the power output of P is 1600 times (= 2.51^8) greater than that of Q. [1] Its power output per unit area is 0.008 (=$(3000/10000)^4$) times that of P. [1] Therefore the area of P is approx 1600 ÷ 0.008 × that of Q = ×200 000 greater [1] so its diameter about 450 × greater than that of Q. [1]

2.4

1   a   The brightness of a star suddenly increases [1] by about 20 magnitudes [1] and then it decreases much more slowly. [1]

   b   Nuclear fusion ceases in a red giant star which has much more mass than the Sun. [1] The core collapses until its protons and electrons are all converted to neutrons and cannot be compressed any further. [1] The collapsing layers outside the core rebound from the neutron core and are thrown into space. [1]

3.1

1   $v = c \Delta\lambda / \lambda = 3.0 \times 10^8 \, m\,s^{-1} \times 0.072\,nm / 620\,nm$ [1]
   = 35 km s^{-1} [1];
   Circumference = vT = 35 km s^{-1} × 2.5 yr × 3.15 × 10^7 s yr^{-1} = 2.76×10^9 km [1]
   so its radius = circumference / 2π = 4.4×10^8 km [1]

3.2/3.3

1   a $m - M = 5 \log \dfrac{d}{10}$ therefore log (**d**/10) = 40/5 [1]
   = 8 [1] hence $d = 10^9$ pc (= 1000 Mpc) [1]

   b   Red shift estimates of the distance to the more distant type 1a supernovae disagree with distance estimates using their apparent magnitude. [1] The discrepancy is explained by assuming they are accelerating. [1]

3.4

1   Depth of the dip / maximum intensity = 0.10. Therefore the area of the planet disc = 0.10 × the area of the star's disc. [1] Since the area of a disc is proportional to the square of its radius, the radius of the planet [1] = 0.32 (=$0.10^{1/2}$) × the radius of the star. [1]

Medical physics

1.1

1   See Figure 1 [1] for each curve (correct shape, peak at about the correct wavelength) [1] for correct relative peak heights

1.2/1.3

1   a   Correctly labelled figure [1]. A diverging (concave) lens of focal length equal to the distance to the eye's far point is placed in front of the eye [1] Parallel rays of light from a distant point object are made to diverge by the correcting lens so they appear to come from the eye's far point [1] and therefore can be focused on the retina. [1]

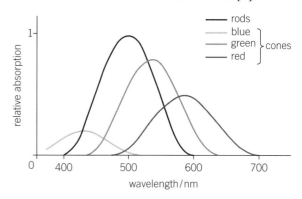

▲ Figure 1

   b $f = -3.00$ m [1] so $P = \dfrac{1}{f} = -0.333$ D [1]

   c $P = \dfrac{1}{f} = \dfrac{1}{+0.25\,m} - \dfrac{1}{0.50\,m}$ [1] = +2.0 D [1]

2.1

1   They amplify the vibrations [1], they filter out unwanted vibrations [1] and they protect the ear from excessive vibrations. [1]

2   The bones of the middle ear act as levers to increase the force of the vibrations [1] and they cause the vibrations to act on a smaller area (the oval window) than the ear drum [1] So the force per unit area (i.e., pressure) they exert on the oval window is greater than the pressure of the sound waves on the ear drum [1]

2.2

1   a $10 \log \dfrac{I}{I_o} = 115$ dB [1]
   therefore $I = 10^{11.5} I_o = 1.0 \times 10^{-0.5}$ W m^{-2} [1] = 0.32 W m^{-2} [1]

   b $10 \log \dfrac{I}{I_o} = 95$ dB gives $I = 10^{9.5} I_o$ [1]. For 2 machines, $I = 2.0 \times 10^{9.5} I_o$ [1]
   so the intensity level = $10 \log (I/I_o) = 10 \log (2.0 \times 10^{9.5})$ = 98 dB [1]

3.1

1   a   i   See Figure 1 [1] for correct shape of figure, [1] for correct scales.

   ii   [1] for correctly positioned labels P, Q, R, S, and T. P is when the atria contract [1]; QRS is when the ventricles contract and the atria relax [1]; T is when the ventricles relax [1].

Answers to summary questions

b Any 2 characteristics from Table 1. [1] for each characteristic, [1] for correct reason

4.1

1 a It absorbs ultrasonic waves created at the rear surface of the disc as these would otherwise cancel each other out. [1] It also damps the disc vibrations rapidly at the end of each pulse before the next pulse is produced. [1]

b An air-skin boundary reflects almost 100% of the incident ultrasonic energy [1]. The probe is applied to the body via a suitable gel or a water bag because most of the ultrasound energy is not reflected and enters the body. [1]

c Acoustic impedances: organ $1.691 \times 10^6 \, kg \, m^{-2} \, s^{-1}$, soft tissue $1.627 \times 10^6 \, kg \, m^{-2} \, s^{-1}$ [1] reflection coefficient 0.019(3) [1]

4.2

1 a The incoherent bundle is used to illuminate the cavity which the endoscope has been inserted into. [1] The coherent bundle is used to see an image of an object inside the body [1] by transmitting light from an image formed by a lens on its end in the body. [1]

b They need to be as thin as possible so the image is as detailed as possible [1] and so the endoscope is flexible and does not lose light where it is bent. [1]

c For the core-cladding boundary, $\sin i_c = \dfrac{1.52}{1.60}$ which gives $i_c = 71.8°$. [1]

Therefore $\sin i_{max} = n_1 \sin(90 - i_c) = 1.60 \times \sin(90 - 71.8)$ [1]

which gives $i_{max} = 30.0°$ [1]

The viewing angle is therefore 60.0°. [1]

4.3

1 a Pulses of radio waves are used in an MR scan to excite hydrogen nuclei [1] systematically across the body using a scanning magnetic field [1]. When the excited nuclei de-excite they emit radio waves [1] that are detected and used to locate the excited nuclei [1] to enable an image of the hydrogen content of the scanned area to be built up. [1]

b The relaxation time of the detected signal is the time the signal takes to decrease. [1] This depends on the type of tissue the excited hydrogen molecules are in [1] and therefore can be used to identify different types of tissue. [1]

5.1

1 a Continuous curve with a peak and tending to zero at long wavelength [1] zero intensity at minimum wavelength [1] intensity spikes [1]

b $\lambda_{MIN} = \dfrac{hc}{eV} = \dfrac{6.63 \times 10^{-34} \, J\,s \times 3.0 \times 10^8 \, m\,s^{-1}}{1.6 \times 10^{-19} \, C \times 60000 \, V}$ [1]

= 0.021 nm [1]

5.2

1 a $\mu = \dfrac{\ln 2}{X_{\frac{1}{2}}} = \dfrac{\ln 2}{0.015 \, m} = 46(.2) \, m^{-1}$ [1] $\mu_m = \mu/\rho =$ 46.2 m^{-1}/1900 $kg\,m^{-3}$

= 0.024 [1] $m^2 \, kg^{-1}$ [1] (answer given to 2 s.f.)

b $I = I_o \, e^{-\mu X_{\frac{1}{2}}} = I_o e^{-(46.2 \, m^{-1} \times 0.004 \, m)}$ [1] $= 0.83 \, I_o$ [1]

Therefore the % reduction of intensity = 0.17 × 100% = 17% [1]

5.3

1 a A collimator grid [1] prevents X-rays scattered in the patient from reaching the film or detector. [1] Scattered X-rays reaching the unexposed parts of film would reduce the contrast between these parts and the exposed parts. [1]
A contrast medium in an organ [1] absorbs X-rays that enter it and prevents X-rays reaching the film [1] so increasing the contrast between the image of the organ and its surroundings [1].

b The intensifying screen is intended to allow the exposure of the patient to X-rays to be reduced without affecting the image. [1] The screen contains a double-sided film between two layers of fluorescent material in a light-proof wrapper. [1] Each X-ray photon causes many light photons to be emitted [1] so affecting the film much more than if the fluorescent materials had not be present. So the exposure to X-rays can be reduced without affecting the image on the film. [1]

6.1

1 a Rearranging $\dfrac{1}{T_{eff}} = \dfrac{1}{T_b} + \dfrac{1}{T_p} =$ gives

$T_{eff} = \left(\dfrac{1}{24} \, hr + \dfrac{1}{12} \, hr\right)^{-1}$ [1] = 8.0 hr [1]

b It is a gamma emitter with a half-life of 6 hours [1] which decays to an isotope which has a very long half-life or can be prepared on site. [1]

Answers to summary questions

6.2

1. **a** Can reach deep tumours [1] less lead shielding needed round the source [1]

2. X-ray photons would need to pass through normal tissues before reaching the tumours whereas the radioisotope attaches itself to the tumour. [1] The X-rays would need to be directed more widely to destroy the tumours whereas the radioisotope targets the tumours. [1]

Engineering physics

1.1

1. **a** Initial angular speed $\omega_o = 2.0 \times \frac{2\pi}{60} = 0.21$ rad s^{-1}

    angular speed after 75.0 s, $\omega = 17.0 \times \frac{2\pi}{60} = 1.78$ rad s^{-1} [1]

    angular acceleration, $\alpha = \frac{(\omega - \omega_o)}{t} = \frac{1.78 - 0.21}{75.0}$
    $= 2.09 \times 10^{-2}$ rad s^{-2} [1]

    b Angular displacement $\theta = \frac{1}{2}(\omega + \omega_o)t = 0.5 \times$
    $(0.21$ rad s^{-1} + 1.78 rad s$^{-1}) \times 75.0$ s [1]
    $= 74.6$ rad [1]

    Number of turns $= 74.6/2\pi = 11.9$ [1]

1.2

1. **a** **i** Angular displacement for 24.6 turns = 24.6 × 2π rad = 155 rad [1]

    Using $\theta = \frac{1}{2}(\omega + \omega_o)t$ with $\omega_o = 0$ gives

    $\omega = \frac{2\theta}{t} = \frac{2 \times 155 \text{ rad}}{38.2 \text{ s}}$
    $= 8.11$ rad s^{-1} [1]

    Its average angular acceleration
    $\alpha_1 = \frac{(\omega - \omega_o)}{t} = \frac{8.11 \text{ rad s}^{-1} - 0}{38.2} = 0.212$ rad s^{-2} [1]

    ii During deceleration, its angular displacement for 97.0 turns = 97.0 × 2π rad = 609 rad [1]
    Its initial angular velocity $\omega_o = 8.11$ rad s^{-1}, its final angular velocity $\omega = 0$
    Therefore its average angular deceleration
    $\alpha_2 = \frac{(\omega - \omega_o)}{t} = \frac{0 - 8.11 \text{ rad s}^{-1}}{92.0}$
    $= -0.0882$ rad s^{-2} [1]

    b During deceleration, the frictional torque $T_F = I\alpha_2 = 290$ kg m^2 × 0.0882 rad s^{-2} = 25.6 N m [1]
    During acceleration, the resultant torque $T = I\alpha_1 = 290$ kg m^2 × 0.212 rad s^{-2} = 61.5 N m

    Therefore the applied torque during acceleration
    = 61.5 N m + 25.6 N m = 87.1 N m [1]
    (Note: the resultant torque = the applied torque − the frictional torque)

1.3

1. **a** 40 J [1]

    b 0.2 J [2]

    c **i** 39.8 J [1]

    ii 1.3×10^{-2} kg m^2 [3]

1.4

1. **a** 0.64 N m s [2]

    b 0.015 kg m^2 [4]

2.1/2.2

1. **a** 180 J into the gas. [1]

    b 180 J of work is done on the gas. [1]

2. **a** 107 kPa, [3] 289 K [2]

    b approx average pressure = 130 kPa, increase of volume = 0.0122 m^3 [1]
    estimated work done = 130 kPa × 0.0122 m^3 [1] ≈ 1.6 kJ [1]
    (Note – use this method to check a more accurate graphical estimate)

2.3

1. **a** 130 kW [2]

    b 32% [3]

2.4

1. **a** 0.73 [2]

    b **i** 31 MJ kg^{-1} × 9.4 kg [1] / 3600 s [1]

    ii 43 kW [1]

2.5

1. **a** 5 [2]

    b 480 W [1]

2. **a** 15.1 kJ [2]

    b 2.5 [2]

Turning points in physics

1.1

1. **a** Every atom contains electrons. [1] The atoms of an element are different from the atoms of any other element. [1] When one or more electrons are removed from an atom, the electrons removed are the same [1] but the atom becomes a positive ion which is therefore different for each type of gas. [1]

    b 2.5×10^7 m s^{-1} [2]

Answers to summary questions

1.2/1.3
1. a 3.85×10^7 m s^{-1} [2]
   b 1.74×10^{11} C kg^{-1} [2]

1.4
1. a 6.20×10^{-15} kg [3]
   b 4.81×10^{-19} C [2]

2.1
1. a When a corpuscles passes from glass into air, its component of velocity parallel to the normal decreases and is component perpendicular to the normal is unchanged. [1] So the direction of its velocity is further from the normal [1] and the light ray therefore is refracted away from the normal. [1]
   b According to Newton's theory, only two bright fringes should be seen corresponding to corpuscles passing through each of the two slits. [1] The fringe pattern always has more than two fringes and Huygens' theory explains this by considering the two slits as coherent emitters of waves. [1] A bright fringe is seen at regularly spaced positions where the waves from each slit arrive in phase [1] and a dark fringe is seen between the bright fringes where the waves arrive out of phase by 180° [1].

2.2
1. a The radio waves from the transmitter are polarised. [1] If the aerial is turned through 90° from its position at maximum signal strength, the strength of the detected signal decreases to zero [1] as the orientation of the aerial relative to the plane of polarisation changes. [1]
   b $c = 4Df_0N = 4 \times 8500$ m $\times 12.0$ Hz $\times 720$ [1] = 2.9×10^8 m s^{-1} [1]

2.3
1. a 4.3×10^{-19} J [2]
   b 9.6×10^{-20} J [3]

2.4
1. a 3.5×10^7 ms^{-1} [2], 2.0×10^{-11} m [2]
   b More detail would be seen as the electrons would be diffracted less at each lens [1] because the electrons would not be slowed down as much by the sample [1] so their de Broglie wavelength would be less. [1]

3.1
1. a The path difference of the light waves from the beam splitter to the observer via each of the two reflectors is an odd number of half wavelengths [1] so the waves are out of phase by 180° on reaching the observer. [1] So the two waves cancel each other and the observer sees a dark fringe. [1]
   b The experimenters expected the fringes to shift. [1]
   c The fringes did not shift so the ether does not exist. [1]

3.2
1. a 96 ns [2]
   b 852 MeV [3]

Electronics

1.1
1. a As the input voltage increases, the gate–source pd increases and when this pd exceeds the MOSFET threshold voltage, a conducting channel between the drain and the source increases [1] so there is a current between the drain and the source and through the relay coil. [1] The relay coil is therefore magnetised and the electromagnet in the relay attracts and closes the relay switch. [1]
   b See Figure 2. Only one curve is required [1] and it should be above the cut-off region [1]. P is left-hand end of the horizontal section of the curve. [1]

1.2
1. a resistor current = (9.0 V – 6.0 V) / 60 W [1]
      = 50 mA [1]
   b diode current = resistor current – lamp current
      = 50 mA – 30 mA = 20 mA [1]

1.3
1. a 3.4 V [2]
   b 6.2 V [2]

1.4
1. a The perpendicular component of the magnetic flux density relative to the device. [1]
   b i 7.1 µT [2]
      ii 45° [1]

2.1
1. a 25 mV [1]
   b 101 001 00 [2]

Answers to summary questions

2.2
1. **a** 2.4 V [2]
   b 9.8 V [2]

3.1
1. **a** 45 pF [2]
   b **The Q factor** is the ratio of the resonant frequency to the bandwidth [1] where the bandwidth is the difference between the frequencies either side of the peak at √2 of the maximum pd. [1]

3.2
1. **a** The output pd of an op-amp saturates if the pd between its two input terminals exceeds about 150 μV. [1] If one of the two inputs of an op-amp is earthed and the output pd is not saturated, the potential at this other input must be less than 150 μV and is said to be at virtual earth. [1]
   b **i** When the buzzer is off, the output pd is at −15 V so the pd at P must be greater than the pd at Q. [1] When the thermistor's temperature decreases, its resistance increases so the pd at P decreases. [1] When this pd decreases below the pd at Q, the output pd switches to +15 V and the buzzer switches on. [1]
   ii The resistance of the variable resistor needs to be increased [1] (so the thermistor temperature does not have to increase as much before the pd at P becomes less than the pd at Q).

3.3
1. **a** See Figure 1 input pd, output pd amplified [1] in phase [1] saturated [1]

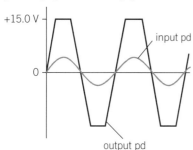

▲ **Figure 1**

   b 6.5 V [2]

4.1
1. **a** [4]

A	B	$\overline{A}$	$\overline{B}$	$\overline{A}\cdot\overline{B}$	$C = \overline{\overline{A}\cdot\overline{B}}$
0	0	1	1	1	0
1	0	0	1	0	1
0	1	1	0	0	1
1	1	0	0	0	1

   b The column for C is the same as A + B. Therefore $\overline{\overline{A}\cdot\overline{B}} = A + B$ [1]

4.2
1. **a** The inputs of a 2-input AND gate are connected to the Q_3 and Q_1 outputs of the binary counter and the AND gate output is connected to the reset terminal of the counter. [1] Each time Q_3 Q_2 Q_1 reaches 0100 which is decimal 4, the next pulse causes Q_3 Q_2 Q_1 to become 0101 [1] so the AND gate output becomes 1 thus resetting the counter to 0000. [1]
   b Each clock pulse causes the Q output of each flip-flop to be shifted to the next Q output. Assuming all the Q outputs are initially 0, the input at the 1st D input is 1 and all the other inputs are zero. [1] The 1st pulse shifts the 1 to the 1st Q output and the second pulse shifts it to the 2nd Q output. [1] There is still a 1 at the 1st D input so this is shifted to the 1st Q output by the second pulse. The third pulse shifts the 2 '1's to the 2nd and 3rd Q outputs and provides another 1 to the 1st Q output so the output states show 1 1 1. [1] Because the 3rd Q output is now 1, the third Q bar output is now 0 so the next three pulses change the 1st Q output to 0 as well as shifting the bits at Q outputs along. The result is the output states count down from 111 to 000. [1]

Input pulse number	Q_1	Q_2	Q_3
0	0	0	0
1	1	0	0
2	1	1	0
3	1	1	1
4	0	1	1
5	0	0	1
6	0	0	0

5.1
1. **a** **i** (HF) radio waves [1]
   ii microwaves [1]
   b 100 [1]; 2.0 Mbps [2]

5.2
1. **a** Bandwidth is the range of frequencies of the signal [1] 97.5 MHz to 98.0 MHz. [1]
   b **i** 85 kHz [1]
   ii 102.45 MHz [1]

For reference

Data

Fundamental constants and values

quantity	symbol	value	units
speed of light in vacuo	c	3.00×10^8	$m\ s^{-1}$
permeability of free space	μ_0	$4\pi \times 10^{-7}$	$H\ m^{-1}$
permittivity of free space	ε_0	8.85×10^{-12}	$F\ m^{-1}$
magnitude of charge of electron	e	1.60×10^{-19}	C
Planck constant	h	6.63×10^{-34}	$J\ s$
gravitational constant	G	6.67×10^{-11}	$N\ m^2\ kg^{-2}$
Avogadro constant	N_A	6.02×10^{23}	mol^{-1}
molar gas constant	R	8.31	$J\ K^{-1}\ mol^{-1}$
Boltzmann constant	k	1.38×10^{-23}	$J\ K^{-1}$
Stefan constant	σ	5.67×10^{-8}	$W\ m^{-2}\ K^{-4}$
Wien constant	α	2.90×10^{-3}	$m\ K$
electron rest mass (equivalent to 5.5×10^{-4} u)	m_e	9.11×10^{-31}	kg
electron charge/mass ratio	$\dfrac{e}{m_e}$	1.76×10^{11}	$C\ kg^{-1}$
proton rest mass (equivalent to 1.007 28 u)	m_p	$1.67(3) \times 10^{-27}$	kg
proton charge/mass ratio	$\dfrac{e}{m_p}$	9.58×10^7	$C\ kg^{-1}$
neutron rest mass (equivalent to 1.008 67 u)	m_n	$1.67(5) \times 10^{-27}$	kg
gravitational field strength	g	9.81	$N\ kg^{-1}$
acceleration due to gravity	g	9.81	$m\ s^{-2}$
atomic mass unit (1 u is equivalent to 931.3 MeV)	u	1.661×10^{-27}	kg

Astronomical data

body	mass/kg	mean radius/m
Sun	1.99×10^{30}	6.96×10^8
Earth	5.98×10^{24}	6.37×10^6

Geometrical equations

arc length = $r\theta$

circumference of circle = $2\pi r$

area of circle = πr^2

surface area of cylinder = $2\pi rh$

volume of cylinder = $\pi r^2 h$

area of sphere = $4\pi r^2$

volume of sphere = $\tfrac{4}{3}\pi r^3$

Momentum

force $\qquad F = \dfrac{\Delta(mv)}{\Delta t}$

impulse $\qquad F\Delta t = \Delta(mv)$

Circular motion

magnitude of angular speed $\quad \omega = \dfrac{v}{r}$

$\omega = 2\pi f$

centripetal acceleration $\quad a = \dfrac{v^2}{r} = \omega^2 r$

centripetal force $\quad F = \dfrac{mv^2}{r} = m\omega^2 r$

Simple harmonic motion

acceleration $\qquad a = -\omega^2 x$

displacement $\qquad x = A\cos(\omega t)$

For reference

speed	$v = \pm \omega\sqrt{A^2 - x^2}$
maximum speed	$v_{max} = \omega A$
maximum acceleration	$a_{max} = \omega^2 A$
for a mass-spring system	$T = 2\pi\sqrt{\dfrac{m}{k}}$
for a simple pendulum	$T = 2\pi\sqrt{\dfrac{l}{g}}$

Gases and thermal physics

gas law	$pV = nRT$
	$pV = NkT$
kinetic theory model	$pV = \dfrac{1}{3}Nm(c_{rms})^2$
kinetic energy of a gas molecule	$\dfrac{1}{2}m(c_{rms})^2 = \dfrac{3}{2}kT = \dfrac{3RT}{2N_A}$
energy to change temperature	$Q = mc\Delta T$
energy to change state	$Q = ml$

Gravitational fields

force between two masses	$F = \dfrac{Gm_1m_2}{r^2}$
magnitude of gravitational field strength	$g = \dfrac{F}{m}$
	$g = \dfrac{GM}{r^2}$
gravitational potential	$\Delta W = m\Delta V$
	$V = -\dfrac{GM}{r}$
	$g = -\dfrac{V}{r}$

Electric fields and capacitors

force between two point charges	$F = \dfrac{Q_1Q_2}{4\pi\varepsilon_0 r^2}$
force on a charge	$F = EQ$
field strength for a uniform field	$E = \dfrac{V}{d}$
field strength for a radial field	$E = \dfrac{Q}{4\pi\varepsilon_0 r^2}$
work done	$\Delta W = Q\Delta V$
electric potential	$V = \dfrac{Q}{4\pi\varepsilon_0 r}$
capacitance	$C = \dfrac{Q}{V}$
	$C = \dfrac{A\varepsilon_0\varepsilon_r}{d}$
decay of charge	$Q = Q_0 e^{\frac{-t}{RC}}$
capacitor charge	$V = V_0(1 - e^{\frac{-t}{RC}})$
time constant	RC
time to halve	$T_{1/2} = 0.69RC$
capacitor energy stored	$E = \dfrac{1}{2}QV = \dfrac{1}{2}CV^2$
	$= \dfrac{1}{2}\dfrac{Q^2}{C}$

Magnetic fields

force on a current	$F = BIl$
force on a moving charge	$F = BQv$
magnetic flux	$\phi = BA$
magnetic flux linkage	$N\phi = BAN\cos\theta$
induced emf	$\varepsilon = N\dfrac{\Delta\phi}{\Delta t}$
emf induced in a rotating coil	$N\phi = BAN\cos\theta$
	$\varepsilon = BAN\omega\sin\omega t$
alternating current	$I_{rms} = \dfrac{I_0}{\sqrt{2}} \quad V_{rms} = \dfrac{V_0}{\sqrt{2}}$
transformer equations	$\dfrac{N_s}{N_p} = \dfrac{V_s}{V_p}$
efficiency	$= \dfrac{I_s V_s}{I_p V_p}$

Nuclear physics

the inverse-square law for γ variation	$I = \dfrac{k}{x^2}$
radioactive decay	$\dfrac{\Delta N}{\Delta t} = -\lambda N \quad N = N_0 e^{-\lambda t}$
activity	$A = \lambda N$
half-life	$T_{1/2} = \dfrac{\ln 2}{\lambda}$
nuclear radius	$R = R_0 A^{\frac{1}{3}}$
energy–mass equation	$E = mc^2$

For reference

Particle physics

Rest energy values

class	name	symbol	rest energy / MeV
photon	photon	γ	0
lepton	neutrino	ν_e	0
		ν_μ	0
	electron	$e^\pm$	0.510 99
	muon	$\mu^\pm$	105.659
mesons	pion	$\pi^\pm$	139.576
		π^0	134.972
	kaon	$K^\pm$	493.821
		K^0	497.762
baryons	proton	p	938.257
	neutron	n	939.551

Properties of quarks
Antiparticles have opposite signs

type	charge	baryon number	strangeness
u	$+\frac{2}{3}e$	$+\frac{1}{3}$	0
d	$-\frac{1}{3}e$	$+\frac{1}{3}$	0
s	$-\frac{1}{3}e$	$+\frac{1}{3}$	-1

Properties of leptons

lepton	lepton number
particles: e^-, ν_e; μ^-, ν_μ	$+1$
antiparticles: e^+, $\bar{\nu}_e$; μ^+, $\bar{\nu}_\mu$	-1

Photons and energy levels

photon energy $\qquad E = hf = \dfrac{hc}{\lambda}$

photoelectric effect $\qquad hf = \phi + E_{K(max)}$

energy levels $\qquad hf = E_1 - E_2$

de Broglie wavelength $\qquad \lambda = \dfrac{h}{p} = \dfrac{h}{mv}$

Waves

wave speed $\qquad c = f\lambda$

period $\qquad f = \dfrac{1}{T}$

first harmonic $\qquad f = \dfrac{1}{2l}\sqrt{\dfrac{T}{\mu}}$

fringe spacing $\qquad w = \dfrac{\lambda D}{s}$

diffraction grating $\qquad d \sin \theta = n\lambda$

refractive index of a substance s $\qquad n_s = \dfrac{c}{c_s}$

For two different substances of refractive indices n_1 and n_2,

law of refraction $\qquad n_1 \sin \theta_1 = n_2 \sin \theta_2$

critical angle $\qquad \sin \theta_c = \dfrac{n_2}{n_1}$ for $n_1 > n_2$

Mechanics

Moments $\qquad$ moment $= Fd$

velocity and acceleration $\qquad v = \dfrac{\Delta s}{\Delta t}$; $\quad a = \dfrac{\Delta v}{\Delta t}$

equations of motion
$$v = u + at$$
$$s = \dfrac{(u+v)t}{2}$$
$$v^2 = u^2 + 2as$$
$$s = ut + \dfrac{1}{2}at^2$$

force $\qquad F = ma$

work, energy, and power $\qquad W = Fs \cos \theta$
$$E_K = \dfrac{1}{2}mv^2$$
$$\Delta E_P = mg\Delta h$$
$$P = \dfrac{W}{t} \quad P = Fv$$

efficiency $= \dfrac{\text{useful output power}}{\text{input power}}$

Materials

density $\qquad \rho = \dfrac{m}{V}$

Hooke's law $\qquad F = k\Delta L$

Young modulus $= \dfrac{\text{tensile stress}}{\text{tensile strain}}$

tensile stress $\qquad = \dfrac{F}{A}$

tensile strain $\qquad = \dfrac{\Delta L}{L}$

energy stored $\qquad E = \dfrac{1}{2}F\Delta L$

Electricity

current and pd $\qquad I = \dfrac{\Delta Q}{\Delta t}$
$$V = \dfrac{W}{Q}$$
$$R = \dfrac{V}{I}$$

emf $\qquad \varepsilon = \dfrac{E}{Q}$
$$\varepsilon = I(R + r)$$

resistors in series $\qquad R = R_1 + R_2 + R_3 + \ldots$

resistors in parallel $\qquad \dfrac{1}{R} = \dfrac{1}{R_1} + \dfrac{1}{R_2} + \dfrac{1}{R_3} + \ldots$

resistivity $\qquad \rho = \dfrac{RA}{L}$

power $\qquad P = VI = I^2R = \dfrac{V^2}{R}$